Classic Geology in Europe 9

The Western Highlands of Scotland

CLASSIC GEOLOGY IN EUROPE

For details of these and other earth sciences titles
from Dunedin Academic Press see
www.dunedinacademicpress.co.uk

Classic Geology in Europe 9

The Western Highlands of Scotland

Con Gillen

DUNEDIN

EDINBURGH ◆ LONDON

Published in the United Kingdom by
Dunedin Academic Press Ltd
Head Office:
Hudson House, 8 Albany Street, Edinburgh, EH1 3QB
London Office:
352 Cromwell Tower, Barbican, London, EC2Y 8NB

ISBN:
9781780460406 (paperback)
9781780466064 (epub)
9781780466071 (Kindle)

British Library Cataloguing in Publication Data
A catalogue record for this book is available from the British Library

Typeset by Makar Publishing Production, Edinburgh
Printed in Poland by Hussar Books

Contents

Acknowledgements

I wish to thank most sincerely those involved with Dunedin Academic Press – Anthony Kinahan, David McLeod, Anne Morton and Sandra Mather – for their patience, support and advice. Anne's editing and proofreading skills helped improve the typescript enormously, and David has done a splendid job with the illustrations and layout, as always. I am also greatly indebted to Professor Graham Park for his comments and for making many valuable suggestions which I happily incorporated; also the comments of an anonymous reviewer.

C. Gillen, Edinburgh, July 2019

Preface

This book presents an introduction to the geology and scenery of the north, north-west and west coasts of Scotland: a region that is arguably the birthplace of geology, and a truly remarkable outdoor laboratory. The region has been intensively studied for 200 years and has provided enormous stimulation to generations of geologists, on account of its variety of rocks and structures in a relatively small and accessible area. Many theories on the origin of mountain belts have been tried and tested here, and the rocks and landscapes continue to attract professionals and amateurs alike.

If you have never visited the area, do come along and be amazed at the views and examine the rocks that gave rise to the magnificent scenery.

The guide includes descriptions of ten of Scotland's 51 best sites for geology: Smoo Cave, Scourie and Loch Laxford, Loch Glencoul, Knockan Crag, Corrieshalloch Gorge, Beinne Eighe and Loch Maree, the Parallel Roads of Glen Roy, Glencoe, Luing and the Slate Islands, and Balmaha and Loch Lomond (see Appendix for details of the scottishgeology.com website).

Frontispiece Ben Arkle and Loch More, Northwest Sutherland. Cambrian Quartzite lying unconformably above Lewisian Gneiss. The quartzite beds are repeated several times by thrust faults parallel to the bedding planes.

Chapter 1

Introduction

This guide is for those who have a passion for the landscape and scenery of northern and western Scotland, and who wish to understand the interplay between rocks and scenery in a truly classic geological region. Nowhere else in Britain is the link more clearly seen.

With stunning mountain and coastal scenery, the region is now very accessible, yet there remains a feeling of remoteness and solitude, making this one of the last wilderness areas of Britain. It is an area of outstanding natural beauty, with two national parks (Cairngorms, and Loch Lomond and Trossachs), many Sites of Special Scientific Interest, designated scenic areas, special areas of conservation; and it hosts Scotland's first two geoparks, the Northwest Highlands and Lochaber. Relations between the various rock types are clearly visible, and the important links between rocks, soils, climate and natural habitats for many rare animals, birds and plants add considerably to the region's conservation value.

When you visit the Western Highlands, you will see some of the oldest rocks in Europe, and a landscape a billion years old that has been exhumed from the depths (Fig. 1.1). Geologists have been examining these rocks since the early part of the nineteenth century and, in spite of the most intense research imaginable in what is just a tiny fraction of the Earth's surface, controversies still surround some of the rock formations. Many fundamentally important concepts in geology were first developed and applied here, then used around the world. The region can fairly be described as a superb natural laboratory for the study of mountain building, including the discovery that huge thick masses of rocks have been turned completely upside down and pushed sideways by over 100km. Representatives of all the major rock types are found here, and their ages span three-quarters of all geological time since the Earth began, 4600 million years ago. The main reason for this incredible variety is related to Scotland's location at the edge of a continent that was caught up in a series of collisions and ruptures,

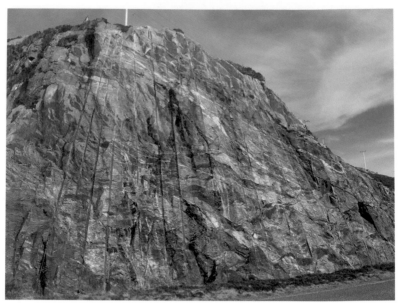

Figure 1.1 Folds and granite sheets in Lewisian gneiss at Loch Laxford, Sutherland; these are among the oldest rocks in the country.

which left their mark in the rocks and structures. That complex story has been unravelled in the past two centuries, based primarily on detailed fieldwork.

Historical background

The earliest records of observations on the landscape features of the Northwest Highlands go back to 1772, when the Welsh naturalist Thomas Pennant (1726–1798), who made important systematic collections of minerals and fossils, published his account of a visit to the district. His book made an immediate impression and helped to establish the unique aspects of Northwest Scotland. Pennant was particularly struck by the unusual mountain shapes, particularly Suilven and other such hills in Assynt, Torridon and Applecross. He also remarked on the marble quarries at Ledmore, near Elphin, and noted the use of marble for making agricultural lime and as high-quality polished stone for statues and decorative slabs. The quarry was in use until 2016.

The father of modern geology, James Hutton (1726–97), was a contemporary of Pennant. He was from Edinburgh and was a leading figure in the Scottish Enlightenment. Although he never worked in the Northwest

Highlands, he was one of the first to use field observations to advance his hypothesis, and was the first to recognize the importance of geological time and the essentially cyclical nature of rockforming processes. As such, he profoundly influenced all who followed him. His *Theory of the Earth* (1788) was a landmark in science publication. The nearest Hutton came to the Northwest Highlands was a visit to Glen Tilt in 1785, when he concluded correctly that the granite on one side of the glen was younger than the adjacent limestone on the other side, for he saw that veins of granite cut and penetrate into the limestone. Up until that time, it had been accepted that granite was the oldest rock in the world, having supposedly crystallized out first from some universal ocean. This is a further example of the importance of the rocks of the Scottish Highlands in the development of the science of geology.

More-systematic investigations began early in the nineteenth century when John MacCulloch (1773–1835), an amateur geologist who later joined the Geological Survey, worked extensively in the area from 1814 to 1824 to prepare the first geological map of Scotland, published in 1836. His endeavours were remarkable, considering the remoteness of the area and the ruggedness of the terrain. He successfully identified and mapped all the major rock units, although their ages and mutual relations were subsequently modified, sometimes quite radically. MacCulloch's results were referred to and built upon for a good half century. In 1819, he also published a beautifully illustrated three-volume account of the geology of the Hebrides – *Description of the western islands of Scotland, including the Isle of Man*. His contribution to the geology of Northwest Scotland cannot be underestimated.

The next major figure to advance the geology of the Northwest Highlands was Sir Roderick Impey Murchison (1792–1871), director of the Geological Survey. Murchison undertook fieldwork from the 1820s onwards. He had previously worked in the Southern Uplands and Wales, and also extensively in Russia, for which he was honoured by the Tsar. With the arrival of Murchison in the Highlands, a famous controversy flared up in the 1860s between survey geologists and academics concerning the overall structure of the rock formations. Murchison, from his work in Wales, assumed that all the rocks became younger upwards, a situation that would normally be expected. However, he did not realize that, in the rocks of north Sutherland, what he thought was bedding was actually formed because of flattening and shearing during Earth movements, and that his 'younger' beds were actually older than those lying beneath. The dispute between Murchison and James Nicol (1810–79), professor at the University of Aberdeen, who maintained that the older rocks were placed above younger by flat-lying faults (thrusts), was long, bitter and highly political. Murchison had Nicol discredited, and then appointed Archibald Geikie (1835–1924) to be director of the Survey in Scotland, then as first professor of geology at the University of Edinburgh.

In the 1880s, Charles Lapworth (1842–1920), a schoolteacher from Galashiels, who had worked extensively in the richly fossiliferous beds of the Southern Uplands, showed by meticulous mapping of the rocks in the Northwest Highlands that the same beds were repeated many times over by folding and thrusting, thus vindicating Nicol. Geikie was then forced to re-examine the area and, between 1883 and 1897, the Northwest Highlands were completely mapped and described in detail by the Survey, principally by Ben Peach, John Horne and Charles Clough, but many others contributed. The shining star in this brilliant galaxy was Clough, the most skilled at mapping. In 1907, a famous memoir was published that has stood the test of time. One colleague of Peach and Horne, Henry Cadell (1860–1934) was the first person (in 1888) to attempt experimenting with models to show how mountains form. He invented a wooden frame in which layers of clay, sand and plaster were squeezed together in a vice to form folds and thrusts (flat faults), and he successfully reproduced structures that he had mapped in detail at Loch Eriboll and Foinaven. A centennial commemorative conference was held in Ullapool in 2007, and a volume of modern contributions reflecting research since 1907 was published in 2010, which in turn has spurred yet more research. The area soon attracted geologists from all over Europe, particularly the Alps, where similar issues were being debated in the 1880s, and the geology of the Western Highlands played a fundamental part in helping to resolve these in another mountain belt.

Farther south, the rocks of the Argyll and the Grampian highlands were also the locus of new ideas, resulting from detailed mapping and interpretation. New techniques were developed, including the use of markers in rocks that indicated how rocks had been completely turned upside down and moved for great distances. This small part of Scotland has inspired and educated geologists from around the world for over 200 years and has been used as a training ground in mapping complex geological structures. Yet in spite of all this intense research, much remains to be done, several hypotheses have to be tested fully, and theories developed and accepted.

Concepts, terms and definitions

Once geology had become established early in the nineteenth century as a science based on deductions made from field observations on how different rocks related to one another, further discoveries led to the rapid advancement of the subject. James Hutton was the first to propose that rocks contained a history of their formation, including evidence for natural processes that had acted over a long period. In particular, the discovery in rocks of fossils that represented various extinct life forms enabled geologists to establish a timetable for the age of rocks, based on the principle that younger rocks rest on older and that modern physical processes (e.g. rivers carry sediment to the sea, where it piles up in layers or beds) acted in the geological past much as they do at present. Rocks were placed in three distinct categories, reflecting the ways in which they formed: sedimentary, igneous and metamorphic.

Geological time

One of James Hutton's most important contributions to science was the recognition that the Earth is immensely old and that it requires millions of years to create changes in rocks by processes that have been continuously shaping and reshaping the Earth since the birth of the planet. These processes have been operating in a cyclical fashion more or less constantly, and in ways that can be observed today. The fundamental concept is that rocks are laid down in sequence, with younger on top of older. Hutton did not have the means available in the 1770s to measure geological time accurately; that breakthrough had to wait another two centuries. However, it was soon realized by careful field observation that fossil remains of animals and plants changed systematically from older rocks to younger ones by forms disappearing through extinction and being replaced by quite new forms in rocks lying above. Thus, a relative timetable was built up, based on the sudden appearance of new fossils. Of course, this method can be used only where fossils occur in sedimentary rocks younger than about 550 million years. Before that time, fossils are sparse and without any solid parts that could be preserved. Indeed, for Scotland the oldest fossils with shells are about 525 million years old. Techniques for dating unfossiliferous sedimentary rocks, and for igneous and metamorphic rocks, had to await the discovery of radioactivity in the 1950s. We now know that the Earth formed 4600 million years ago. The immensely long timespan until the Cambrian Period with the great explosion of life (i.e. four billion years) is referred to as the Precambrian. Subdivisions of the Precambrian are made on the basis of radiometric ages on key rocks, a method that is radically different from the relative method based on dating fossils. The study of fossiliferous sedimentary rocks is called stratigraphy (Latin: stratum, a layer or bed), and the geological timetable that emerged and is used worldwide is referred to as the stratigraphic column, developed in England by the pioneering geologist William Smith (1769–1839), the father of English geology, who produced the first geological map of England based on the principles of stratigraphy (he was nicknamed Strata Smith). Table 1.1 shows a version of the stratigraphic column appropriate for the north of Scotland. The subdivisions are used around the world on all geological maps, and the age of the base of each period has been agreed at an international level. Terms such as Lewisian, Torridonian, Moine, Dalradian, etc. are local and refer only to Scotland.

Geological setting

Since 1970, geologists have known that the Earth's crust is divided into large plates that move slowly around the globe, colliding together to form larger plates, and splitting into smaller plates. In the process, collisions lead to the amalgamation of smaller fragments and the formation of supercontinents separated by large oceans. Cycles of supercontinent formation and destruction typically take 400–500 million years to complete.

Northwest Scotland has some of the oldest rocks in Europe, the Lewisian gneiss, dated at 2900 million years old. The Lewisian formed part of the ancient crust of a large continent called Laurentia (North America and Greenland, effectively). Laurentia in turn was part of the much larger supercontinent of Rodinia (Russian: motherland) that began to split apart 1200 million years ago by rifting and the formation of a new ocean: the Iapetus Ocean. The various rifted segments drifted away to opposite sides of the Iapetus. Rivers from the west brought sand and pebbles, from the now exposed Lewisian rocks, to form the Torridonian. Here we have the first life-forms in Scotland, in the shape of algal mud mounds (called

Table 1.1: Geological timetable for the Highlands, showing names of geological periods and the age of the base of each period in millions of years (m.y.). Oldest periods are conventionally shown at the base of the table.

Geological Period	Age of base in m.y.	Main geological events in Northern Scotland
Quaternary**	2	Sudden climate cooling: Ice Age – erosion by glaciers, then deposition of sediment by rivers from melting ice 10,000 years ago; sculpting of modern landscape
Neogene*	23	Extensive erosion by rivers, wet sub-tropical climate
Palaeogene*	65	Volcanoes in Ardnamurchan and Inner Hebrides; eastward tilting of land surface and formation of main east-flowing river pattern; deep weathering begins
Cretaceous	146	Very hot climate, deposition of Chalk deposits
Jurassic	200	Warm shallow seas; formation of limestone, sandstone, clay and shale with abundant fossils
Triassic	250	New Red Sandstone desert deposits (sand dunes and pebble beds from flash floods). Scotland just north of the Equator
Permian	300	
Carboniferous	360	Sandstones formed by rivers; Scotland on the Equator. Coal swamps in Central Scotland
Devonian	416	Erosion of high Caledonian mountains and formation of Old Red Sandstone continental redbeds (in desert conditions)
Silurian	444	End of Caledonian movements; Moine Thrust, Highland Boundary, Great Glen and Southern Upland faults formed; intrusion of granites in Highlands
Ordovician	488	Durness Limestone formed in shallow tropical seas. Folding and metamorphism of Dalradian rocks, Caledonian mountain building begins
Cambrian	542	Sandstone formed on wide beaches; first shelly fossils
Proterozoic	1000	Torridonian (Northwest Highlands); Moine (Northern Highlands); Dalradian (Central and Grampian Highlands)
Archaean	3000	Lewisian Gneiss – oldest rocks form basement in the Northwest Highlands, other rocks deposited on top

* Palaeogene and Neogene are sometimes referred to informally as 'Tertiary'
** Palaeogene, Neogene and Quaternary are grouped together as the Cenozoic

stromatolites) that grew on the floor of shallow lagoons. Finer material was carried out to sea and piled up to form what was to become the Moine and younger Dalradian rocks in offshore basins. Shallow seas eventually emerged in what is now Northwest Scotland, and clean white sand was laid down on the beaches of an extensive coast, under very stable conditions. At that time, 500 million years ago, Scotland lay south of the Equator within the tropical zone. The white sands were replaced by shallow-water limestone (Durness Group) that formed in bays and lagoons, and some of the earliest animals with shells were preserved as fossils.

About 250 million years after separating, the various continental blocks began to move towards one another in a different configuration that resulted in the Iapetus Ocean being consumed by subduction and the continents colliding. First, Laurentia (with Scotland) collided with Baltica (Scandinavia) 480 million years ago, then these two collided with Avalonia (Europe with England and Wales) 50 million years later, at 430 million years ago. Taken together, these events constitute the mountain-building episodes known as the Caledonian orogeny, during which the Moine and Dalradian rocks were folded and metamorphosed by heat and pressure into schist and slate (Fig. 1.2). The final event was the thrusting westwards of Moine rocks over the Lewisian basement, and movements on the Great Glen Fault, the Highland Boundary Fault and the Southern Upland Fault, to produce Scotland more or less in its present configuration (Figs 1.3, 1.4). It is these Caledonian faultlines that have created the block structure of the country – the Hebrides and Northwest Highlands west of the Moine Thrust, the Northern Highlands from there to the Great Glen Fault, the Grampian and Southwest Highlands as far as the Highland Boundary Fault, the Midland Valley north of the Southern Upland Fault, and the Southern Uplands to the south, as far as the boundary with the Lake District and Cheviot Hills, known as the Iapetus Suture. The suture marks the collision zone between the two continental plates mentioned above (Fig. 1.5).

With the Iapetus Ocean now closed, Scotland found itself in the interior of a large continent, Laurasia (Laurentia + Asia), just south of the Equator and above sea level. The mountainous landscape, devoid of plants and

Figure 1.2 Buachaille Etive Mòr near the entrance to Glencoe, part of a huge igneous complex of Devonian age.

Figure 1.3 Loch Lomond and the Highland Boundary Fault, marked by the line of islands in the loch and continuing to Conic Hill on the right.

Figure 1.4 Loch Glencoul and the world-famous view of the Glencoul thrust: Lewisian at the top has been thrust over younger quartzite, which in turn lies unconformably on Lewisian at the loch shore; Stack of Glencoul on the right, with Moine rocks above the Moine Thrust.

soils, was rapidly downwasted, and the sediments stripped off the mountains in flash floods were carried down to the desert floor to create the Old Red Sandstone. Scotland drifted northwards but remained largely above sea level for the next 350 million years, until the supercontinent Pangaea (Greek: the whole Earth) began to rift and break up, with the formation of the present Atlantic Ocean. The rifting started in the south and gradually spread north, with seas forming around Scotland. By 60 million years ago,

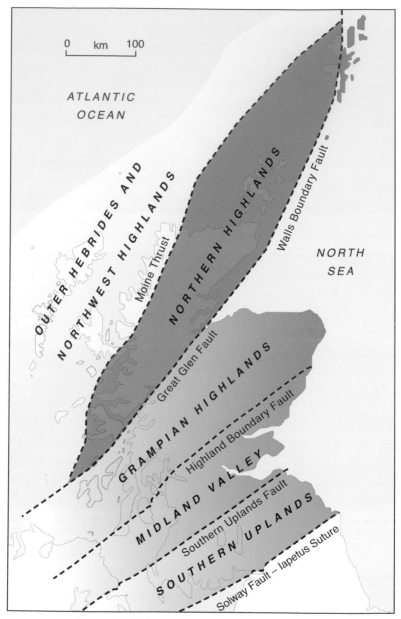

Figure 1.5 Main structural blocks (terranes) of Scotland.

the rift had reached the northern regions of the ocean, and the **intrusion** of a huge plume of molten **magma** originating in the upper **mantle** gave rise to the Hebridean volcanic islands. This plume bulged up the crust, and the mountains of the west coast mainland opposite the volcanic chain were elevated to renewed heights. For 50 million years, the near-tropical climate caused the mountains to be worn down by chemical weathering and physical erosion, with much of the material transported to the Minches and the North Sea.

By two million years ago, Scotland had drifted to its final northerly latitudes in a period of global cooling. Erosion by thick moving continental ice sheets and mountain glaciers removed the last vestiges of the Neogene (late Tertiary) weathered ground, as well as producing an overall smoothing of the landscape. As the ice wasted away 10,000 years ago following a sudden climatic warming, huge rivers deposited thick layers of boulders, gravel, sand and clay in valleys across the country. During the glacial period, the 2km thick mass of ice weighed down on the land surface and caused it to be depressed temporarily. Once the ice disappeared, sea level rose, then gradually the land regained its previous elevation, leaving beaches and cliff lines stranded inland, marking the previous edge of the sea. Such raised beaches and rock platforms are common on the west coast and around the islands. When the Atlantic waters cooled at the start of the ice age, many planktonic organisms were suddenly killed off and their tiny carbonate shells fell to the bottom to form shell banks. Onshore currents and winds have been responsible for transporting this shell sand to the west coast, where it has been shaped into the machair beaches and dunes, a unique feature of the coastal landscape. The machair is found only in Northwest Scotland, Orkney, Shetland and Ireland, and is protected because of its important landscape and wildlife value. Geological processes continue today, of course, and some of the results include landslips, rockfalls with scree, and the reshaping of the coastline.

General information

Anyone who has not yet visited the Highlands of Scotland can expect to be surprised and delighted by the beauty and variety of the landscape, and by the fascinating geology that is responsible for this great diversity. Although the region can be reached quite quickly by road from Glasgow, Edinburgh, Inverness and Aberdeen, progress on narrow twisting roads once there can be slow, especially at the height of the tourist season, and on single-track roads. Each of the routes will take at least a week to complete, some considerably more if every locality is included. References are given to other excursion guides, to allow for more complete (and more technically advanced) coverage. It should be borne in mind that safety is paramount, both when driving and walking. In particular, before beginning any trip, a safe offroad lay-by should be found. Do not be distracted by looking closely at rocks and scenery while driving, and never park in passing places.

This guide is primarily aimed at small groups of people travelling by car. Parking for large coaches is extremely limited at most localities, and some are simply inaccessible by coach. It goes without saying that the country code should be followed at all times (use footpaths, close gates, remove all rubbish, do not disturb farm animals or wildlife); and, if venturing on any of the longer walks, it is advisable to leave a note of the route and expected return time at your accommodation.

Given the sparse nature of the population, facilities such as petrol stations, shops, banks, cash machines, post offices and public toilets are widely scattered, and opening hours, particularly in the evenings and on Sundays, are greatly reduced in comparison to those in large towns. Petrol consumption on Highland roads is surprisingly high (as are prices). Hotels, bed and breakfast establishments, youth hostels and caravan–camping sites are full in summer, making advanced booking essential. Mobile telephone reception can be unreliable, and telephone kiosks are far apart, and are gradually being removed because of lack of use. Individual chapters contain information about local accommodation and visitor facilities. Websites are included at the end of the book, to assist in booking and forward planning.

A word about the notorious Highland midge: beware! They can turn a pleasant excursion into a miserable experience, and some protection is advisable. Spring and late autumn are midge-free periods. Local shops stock various remedies. Ticks from sheep and deer are widespread in long grass and bracken, so protection is advisable (i.e. no shorts or short sleeves). Water quality cannot be guaranteed, owing to sheep and fish-farming pollution, so it is essential to carry bottled water. If dogs are taken, they must be kept under strict control, especially during the lambing season. Many estates do not allow dogs at any time.

Route planning

Because of the relative remoteness of some of the areas and the distances between each locality on the excursions, it is not practical to attempt these excursions using public transport. Although each route requires a full week, it would be possible to select the highlights and spend two weeks from Tongue to Assynt, then Ullapool to Fort William, say. This would then provide a flavour of the tremendous variety of rocks that could be studied in detail on subsequent trips. The best period for fieldwork is early summer, but in fact the period from Easter until the end of September is entirely suitable. Midges are at their worst from June to August, especially on calm, warm, humid days in the morning and evening. Rain can be expected at any time, so waterproofs should be carried.

Most of the localities can be reached on foot by walking a short distance from a safe parking place, but some longer walks are also included. These tend to be around the coast, and most are not tide dependent. A few walks go across higher ground without clear footpaths, and sometimes across peaty ground, which will be wet after rain. For the more remote walks, detailed Ordnance Survey maps are essential, as is a compass, although GPS navigation aids can be helpful (but not for giving elevation). A good initial impression of the terrain can be obtained from Google Earth. Figure 1.6 shows the location of the excursions, overlain on a simplified sketch map of the geology.

Safety in the field

These excursions have been designed to be safe and easily accessible, although any outdoor pursuit carries a certain amount of risk. Common sense and the observance of a few basic rules are all that are required to ensure an enjoyable and trouble-free trip. Most accidents in geology happen when people are driving to the field area, so please take great care not to be distracted by the scenery on the narrow roads and to park safely. When driving on single-track roads, use passing places to allow overtaking. Wear good stout boots and thick socks for comfort, and carry waterproofs and spare clothes in a rucksack on the longer walks. Food, water, a small first-aid kit, torch, whistle and thermal blanket (weightless) will be required on the longer walks, which could last a full day. A safety helmet should be worn when working beneath cliffs and overhangs. A fluorescent yellow overjacket is good when walking along roads, as there are no pedestrian footways in the Highlands. Care should be taken when parking, as there are often hidden (and flooded) ditches and soft verges at the roadside. As well as personal safety, it is equally important to look out for others when doing fieldwork. Take care not to dislodge blocks that might fall.

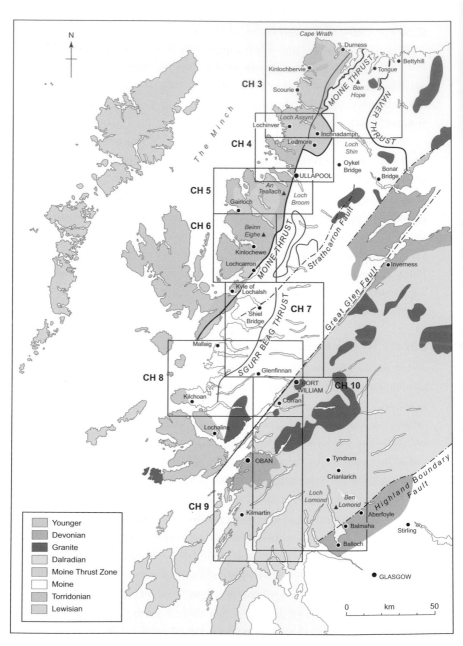

Figure 1.6 Map of geology and excursions.

Stop to write notes in a notebook and do not step backwards when taking photographs. Geology in the field is a pleasant activity that is perfectly safe if these precautions are followed.

Since a key attraction is also the landscape and scenery, it is not advisable to venture out in bad weather, when there will be no vistas to admire. Avoid coastal routes in stormy and misty weather, for many rock slabs dip towards the sea and become very slippery. This is also the case near rivers and waterfalls. Specific advice is given in each chapter, appropriate to the particular localities. Special safety advice relates to mountain climbing and hillwalking, and this is referred to where appropriate. All the excursions in this book are low-level hillwalks and coastal routes, many of which are on clearly marked footpaths. An exception is the walk to the summit of Ben Nevis. Use paths where they exist, and do not cause further erosion by straying off the paths.

Collecting samples

This guide is not a rock collector's guide. Many of the sites are used for teaching purposes, and some are Sites of Special Scientific Interest (SSSI) where collecting is restricted. However, visitors may wish to take home some attractive rocks and minerals, and a few hints are presented here. First, try to take loose material left behind by other collectors rather than remove more material from a rockface. Samples should be trimmed to remove sharp edges and corners, and they should then be numbered and noted in a field notebook with location (place name and grid reference), date and features of interest recorded. Wrap the samples in newspaper. There are only a very few fossil localities mentioned in this book, and readers are referred to the Scottish Natural Heritage (SNH) Fossil Code (2008) for details on how to collect and curate fossils. A handlens is needed to look at minerals, rocks and fossils in the field, either ×8 or ×10 magnification. When hammering rock outcrops, take note of the safety precautions mentioned previously. At all costs, please avoid defacing attractive features such as folds, narrow shears and contacts between different rocks, especially dykes and country rocks. The text makes clear those sites where all collecting is prohibited, including loose material. Far too many localities have been totally ruined by indiscriminate hammering and drilling. Please do take geological conservation seriously, and enjoy the sites, leaving them for others to enjoy later, for they have a very important role to play in teaching the next generation of geologists. Make good use of a field notebook to record details of photographs (i.e. date, location, grid reference and features of interest). Bear in mind, too, that a rucksack full of waterproofs, spare clothes, water, food, hammer and samples will be quite heavy.

Maps

The maps in this book are highly simplified sketchmaps, to illustrate locations, routes and very general geology. For detailed work, it will be necessary to purchase Ordnance Survey (OS) topographic maps at 1:50 000 scale (the Landranger series), or 1:25 000 scale (Explorer series) for greater detail, but at greater cost. These can be purchased in digital form, for use on mobile devices. British Geological Survey (BGS) maps at 1:50 000 scale are available for most of the localities; these can be viewed on a mobile phone or tablet. Note that they are not suitable for navigation: the geology is shown in dense colour on top of older topographic maps, so that hill contours, place-names, roads, rivers, and so on, have been effectively obscured, hence the need to carry two different maps. The BGS website has a map portal, and it is possible to view all their published maps online. Useful summaries of the geology are printed in the map margins. When venturing far from the road, it is essential to use a map and compass (or GPS instrument) properly. Help and advice in this regard can be found on the OS website. Please note that GPS instruments are not reliable for altitude; in addition, satellite coverage can be very patchy at times. If a GPS instrument is to be used frequently, spare batteries should be carried. Each chapter lists the appropriate maps. Another series at 1:250 000 scale is published by the BGS, but they are very general and contain almost no topographic information, and are of limited use. For a good overview, the fifth edition of the BGS map of Scotland (2008) is worth purchasing. This is at the scale of 1:625 000 and includes a comprehensive explanatory booklet at a reasonable price. There is also a 1:1 250 000 map of the whole of the British Isles (2017), but this has less detail. In this book, eight-figure grid references are used, which means that localities are accurate to within 10m. Grid references are printed in square brackets, with two letters indicating the 100km national grid square, followed by four digits for the easting (vertical lines, left to right) then a space followed by a further four digits for the northing (horizontal lines, bottom to top). Thus, for example, the parking place in Scourie Bay is quoted as [NC 1511 4458]. Further help is provided in the margin of OS maps.

Chapter 2

Geological evolution of the Highlands

Scotland can be divided geologically into five large blocks (also called terranes), separated by major faults (Fig. 1.5). This division is a result of earth
movements during the Caledonian orogeny some 470–400 million years
ago. The distinct blocks are the Outer Hebrides and Northwest Highlands, bounded by the Moine Thrust that runs from Loch Eriboll in the
north to Skye in the south. East of this fault lie the Northern Highlands,
which extend to the Great Glen Fault. Between here and the Highland
Boundary Fault are the Grampian and Argyll highlands, with the Midland
Valley to the south, as far as the Southern Upland Fault. The Southern
Uplands extend south towards the Cheviot Hills and the Lake District,
the boundary being known as the Iapetus Suture – a large structure that
formed when the former Iapetus Ocean closed, resulting in the collision
of what were once two separate plates of the Earth's crust.

In the far northwest of the country are the oldest rocks, up to 3000
million years old, the Lewisian gneiss, named after the Isle of Lewis, and
forming the basement (Fig. 2.1). The Outer Hebrides are made almost
entirely of this ancient metamorphic rock, which is also found on the
west coast mainland. Lying above the gneisses are the Torridonian rocks
– mainly red sandstones – which make up some of the striking mountains
of the northwest seaboard, such as Suilven, Canisp and Cùl Mòr (Fig. 2.2).
These rocks were laid down by rivers 1200–900 million years ago, directly
on top of the basement gneiss. Overlying the Torridonian rocks are the
sparkling white Cambrian quartzites of Arkle and Foinaven (Fig. 2.3), followed by the Durness carbonate rocks (limestone), beautifully exposed at
the inlet of Smoo Cave near Durness, and responsible for the green and
fertile fields around Elphin.

The Moine Thrust zone forms an important structural break, marking
the western limit of the Caledonian mountains. Rocks formed during these
earth movements were carried west for a distance of at least 150km by a

Figure 2.1 Cape Wrath, Lewisian gneiss cliffs, showing recumbent folds.

Figure 2.2 Suilven, Canisp and Cùl Mòr from Stac Pollaidh: isolated mountains resting on a basement of Lewisian gneiss.

series of thrust faults that were active 435 million years ago. Much of the mountainous terrain of the Northern Highlands east of the thrust is made of Moine rocks, a series of schists and quartzites about 870–1000 million years old. Originally, these were sands carried by rivers and deposited in shallow sea basins. Moine rocks were folded, metamorphosed and intruded by granitic bodies around 850 million years ago.

Figure 2.3 Cambrian Quartzite screes above Lewisian gneiss on Foinaven, Sutherland.

Younger schists, plus a wide range of other metamorphic rocks, form the Dalradian sequence between the Great Glen and Highland Boundary faults. These were originally deposited as marine sediments 730–470 million years ago, then folded and metamorphosed during the Caledonian mountain-building events that began 470 million years ago and ended with faulting, thrusting and granite intrusion during the period 435–400 million years ago. The Caledonian events also affected the Moine rocks in the Northern Highlands.

Around the edges of the Highlands are younger sedimentary rocks, mainly the Old Red Sandstone of Devonian age, and the New Red Sandstone of Permian to Triassic age, followed by Jurassic shales, mudstones and limestones. Also of Devonian age are the Lorn lavas in western Argyll, around Oban and Kerrera. At Lochaline there is a small area of Cretaceous sandstone that is still actively mined for use as a glass sand.

The youngest rocks are found on the west coast, in the Inner Hebrides – Skye, Rum and Mull in particular – and the Ardnamurchan peninsula. These are the volcanic and intrusive igneous rocks of the Palaeogene (early Tertiary) period, 60–55 million years old. During that event, related to the opening of the North Atlantic Ocean, large volumes of magma were injected into the base of the crust, resulting in uplift of western Scotland and a consequent tilting of the land surface to the east. Much lava was poured out from volcanoes over the surface of Triassic and Jurassic rocks. Intense erosion then occurred during the next 50 million years, and vast amounts of material were removed by rivers in a mild and wet climate. During the Quaternary period, about two million years ago, the climate cooled quite suddenly, and Scotland and the rest of the northern hemisphere experienced a number of glaciations, when the entire country was under ice. Moving glaciers carved the mountain sides and deepened the valleys, and deposition of sand and gravel from meltwater streams at the end of the most recent ice age formed a thick cover of unconsolidated sediment on lower ground. After the ice melted 10,000 years ago, many upland areas were subjected to avalanches and landslips, the results of which are still visible. Action by rivers and the sea continues to shape the landscape.

Lewisian

Lewisian rocks are among the oldest in Europe. The name was first used by Sir Roderick Murchison. Ben Peach and John Horne described the rocks as belonging to a 'fundamental complex', made mainly of metamorphosed igneous and minor amounts of sedimentary rocks. These were then folded together several times and cut by swarms of basic dykes, then all of these were folded and metamorphosed again, until finally being intruded by granite sheets and pegmatite dykes. Thus, the Lewisian complex does not resemble younger sedimentary formations, in which bedding, fossils and other original or primary features are still preserved. Quite different techniques are therefore required to investigate the gneisses and unravel their complex history. Lewisian rocks are found throughout the length of the Outer Hebrides, on the mainland west of the Moine Thrust from Cape Wrath in the north, through Assynt to southern Skye, and on the Inner Hebridean islands of Coll and Tiree. There are smaller occurrences on Iona, Rum, Raasay and Rona. East of the Moine Thrust, there are isolated strips of Lewisian gneiss folded in with the Moine rocks of Sutherland, and it is very likely that the Lewisian underlies the whole of the Northern Highlands, forming the basement to the Moine rocks.

Since the pioneering work of Peach and Horne, who published their memoir in 1907, the Lewisian area has attracted hundreds of researchers, all of whom pay tribute to the quality of the original mapping carried out by Peach and Horne for the Geological Survey. It was they who divided the mainland Lewisian into Northern, Central and Southern regions, each region having its own distinctive rock types and geological history. The Northern region stretches from Durness to Loch Laxford and consists predominantly of pale banded gneisses (white and pink quartz and feldspar alternating with black and dark green biotite and hornblende) with sheets of metamorphosed basic dykes (originally dolerite and basalt, now black hornblende–biotite schist and amphibolite) and many dykes, sheets and irregular patches of pink granite and coarse-grained pegmatite (Fig. 2.4). Along the prominent NW–SE coastal indentation of Loch Laxford is the Laxford shear zone, a 5km-wide belt of vertical and intensely deformed rocks that marks the boundary between the Northern and Central regions. This zone can be traced as far as Ben Stack (721m), one of the highest hills of Lewisian rocks on the mainland. A proposed revision of the naming of Lewisian blocks in 2000 resulted in the Northern region being referred to as the Rhiconich terrane, after the village of Rhiconich, north of Loch Laxford. Not all research workers accept the new interpretation, and Lewisian geology still continues to provoke heated debate, in spite of intensive age-determination efforts.

The Central region extends from Loch Laxford to Gruinard Bay and includes the area around Scourie, the type area for the foreland complex and the Scourian gneisses. In the 1950s the husband and wife team of Janet Watson and John Sutton (from Imperial College London) remapped the

Figure 2.4 Cnoc Gorm, north of Scourie: basic Lewisian gneiss (gabbro), cut by quartz–feldspar pegmatite sheets; in the Laxford shear zone.

areas around Scourie–Laxford and Loch Torridon, and developed the notion that the Lewisian complex consisted of Scourian rocks that were folded and metamorphosed in the Scourian orogeny, then intruded by a single swarm of basic dykes (the Scourie dykes), then subsequently the gneisses and dykes together were modified and reworked by metamorphism and folding during the later Laxfordian orogeny. The dykes were assumed to have been intruded in a single event that was therefore taken to be a time marker, separating the Scourian and Laxfordian events. This interpretation remained current, albeit with local refinements based on age dates, for the best part of half a century, although the assertion of a single swarm of dykes was challenged by several workers, including Donald Bowes from the University of Glasgow, who demonstrated that there were several different sets of dykes, with varying compositions, trends and internal structures, indicating several swarms, thereby confronting the working hypothesis that rocks and structures cut by dykes were 'Scourian' and those showing folded dykes had to be 'Laxfordian'. Age determinations on dykes are very difficult to obtain, but most published dates are 2400 million years, with a few at 2000 million years (see Davies & Heaman, 2014 for details of age dates on Scourie dykes near Lochinver; and Baker *et al.*, 2019 for the younger dates farther south). Dyke swarms are normally seen as short-lived events of 1–5 million years duration, so a difference of 400 million years is implausible as a single event. In addition, the original 1907 survey maps of the Northwest Highlands clearly show at least four different trends and compositions for dykes in the Lewisian. However, this was the first use of dykes as time markers in an attempt at unravelling complex tectonic and metamorphic sequences.

Modifications to the Sutton and Watson scheme included renaming the Scourian the Badcallian (after Badcall village, south of Scourie), and the addition of the Inverian episode, following the discovery of shear zones at Lochinver in the gneisses that pre-date the Scourie dykes (Tarney, 1963). More significant changes were proposed in the period 2000–2005, as will be explained below.

In the Central region, the Scourian gneisses are generally regarded as having been little affected by Laxfordian events. The dykes are nearly vertical and they trend NW–SE. Although metamorphosed, they still retain clear igneous textures, especially away from their edges, where shearing has caused narrow margins of hornblende schist due to metamorphism of the original basalt. Within the Central region, the rocks are mainly dark grey and black coarse-grained banded gneisses, with a gentle dip. The gneisses were formed 3000 million years ago as melts from the upper mantle. Many bodies of black basic and ultrabasic igneous rocks are found as layers, sheets, balls, lenses and flattened streaks within the gneisses and parallel to the gneiss structure. These bodies, which vary in size from many hundreds of metres to just a few centimetres, are older than the basic dykes and in some publications (including the original 1907 Memoir) they are referred to as 'early basic bodies'. Many of the larger basic–ultrabasic bodies show distinct mineral layers of pyroxene, olivine and feldspar that likely formed by gravity settling of heavier crystals during cooling of large igneous intrusions that were flattened and broken up during later folding. In detail, the larger bodies of layered rocks have a different chemical composition from the small dark streaks and pods that are present almost everywhere throughout the gneisses. The layered bodies crystallized from a magma chamber in the crust, whereas the pods crystallized from the original material that gave rise to the quartz–feldspar gneisses of the country rocks (this is explained in papers by Rollinson & Gravestock, 2012, and Guice et al., 2018). Closely associated with the layered bodies are small patches of mica schist and quartzite that formed initially as sedimentary layers on the ancient crust (originally as shale and sandstone respectively). These are referred to as metasediments, and they became interleaved with the basic–ultrabasic igneous rocks during folding and thrusting. Most of the rocks in the Central region are described as granulites – high-grade metamorphic rocks with even-sized coarse crystals making a granular texture, that were deformed at depths of more than 35km in the crust and at temperatures of 900°C or more at the peak of metamorphism during the Badcallian event 2600 million years ago.

Distinctive features of the age of the original rocks and their metamorphism in the northern and southern parts of the Central region has resulted in the subdivision of the region into the Assynt terrane from Loch Laxford to Loch Broom, and the Gruinard terrane from Loch Broom to Gairloch. The age of the original rocks that form the Assynt terrane is 3000 million years, and they were affected by high-grade metamorphism at 2600 million years ago. The Assynt and Rhiconich terranes both show evidence for a later metamorphism at 1750 million years ago, which is also the age of many pegmatite dykes and granite sheets.

The Southern region, the boundary of which runs through Little Loch Broom, includes the Lewisian rocks from Gruinard Bay to Gairloch, Torridon, Rona and Raasay. It resembles the Northern region in that the gneisses have been strongly affected by Laxfordian folding and metamorphism. The boundary between the central and southern regions is an 8km-wide shear zone between Gruinard Bay and Loch Maree. An important group of metamorphosed sedimentary and volcanic rocks occurs in the southern region at Gairloch and again on the northeastern shores of Loch Maree, beneath the Torridonian mountain of Slioch. These make up the Loch Maree Group, which formed after the metamorphism of the Scourian rocks and after the intrusion of the Scourie dykes. The Loch Maree Group was formed 2000 million years ago as a sequence of sedimentary rocks (greywacke, sandstone, shale, mudstone, ironstone and limestone), with basic lavas (basalt) and intrusive igneous sills (dolerite). Deformation and metamorphism of these original rocks produced various gneisses and schists, together with quartzite, marble and amphibolite. At Loch Maree, the rocks are found within a tight, narrow synclinal fold with

Table 2.1: Geological history of the Lewisian complex.

Time, m.y. ago	Geological events and rock types produced
3450	Oldest known components – inherited zircons in gneisses
3300–3000	Formation of original Scourian volcanic and sedimentary rocks
3000	Intrusion of igneous bodies, metamorphism to form gneisses of Central region (Assynt terrane)
2800	Formation of gneisses from sedimentary and volcanic rocks, in Northern region (Rhiconich terrane) north of Loch Laxford
2700	Intrusion of layered basic and ultrabasic igneous bodies at Scourie and Gruinard Bay
2600	Major metamorphic event, high pressure (35–40km deep) and high temperature (900°C), intense deformation: Badcallian event
2500	Folding, metamorphism, shear belts: Inverian event (Badcallian and Inverian were know collectively as the Scourian, in the Central region or Assynt terrane)
2400	Intrusion of dolerite dykes, WNW trend – Scourie dykes
2000	Additional sets of dykes intruded, of various compositions and trends. Formation of Loch Maree Group (lavas and sedimentary rocks)
1850	Folding, metamorphism, shear zones: Laxfordian events; folding of Scourie dykes and metamorphism to amphibolite (hornblende schist); intrusion of granite sheets and pegmatite dykes at Loch Laxford
1750	Metamorphism; formation of Laxford shear zone and amalgamation of Northern and Central regions (Rhiconich and Assynt terranes). Pegmatite dykes and granite sheets intruded
1400–1200	Uplift of Lewisian basement, erosion and exposure at the surface, deep weathering, followed by deposition of Torridonian sedimentary rocks above a major unconformity

an axis trending NW–SE. Studies of the age and chemistry of the Loch Maree Group indicate that one unit formed as an accretionary prism (a wedge of sediments) at the junction between continental and oceanic crust, which were then involved in a continent–continent collision that led to folding, metamorphism and crustal thickening at 1860 million years ago. The other unit consists of metamorphosed ocean-floor basalt lavas, now converted to amphibolites, and the two units were brought together (or accreted) as slices of crust during tectonic movements. Taken together, the basalt and metasediment pair forms a subduction–accretion complex. This is very well described by Park *et al.* (2001).

In both the Northern and Southern regions, there are many pink and red granite sheets, a particularly intense concentration of which is found at Loch Laxford. These were mostly intruded as dykes and veins 1850 million years ago.

Torridonian

Torridonian rocks occur on the northwest coast between Cape Wrath and Rum, and make up part of the sea bed in the Minches and Sea of the Hebrides as far south as Iona (Figs 2.5, 2.6). The rocks were deposited as conglomerate, sandstone, shale and mudstone by rivers flowing mostly

Figure 2.5 Handa Island, showing gently dipping landscape of Torridonian rocks, opposite Scourie (Lewisian gneiss in foreground).

Figure 2.6 Alligin, Wester Ross, illustrating thick accumulation of Torridonian sandstone.

eastwards, at the edge of the Laurentian continent in the period 1200–950 million years ago. The Torridonian sandstones form one of the major rock units in Britain, and occupy a volume of at least 150 000km^3. Typically, they are dark red and brown coarse sandstones, with very gentle dips, having mostly avoided the Caledonian earth movements, except in the Sleat district of Skye, where they were involved in the Moine Thrust complexities. Torridonian mountains are among the most majestic in the country, and they give the Northwest Highlands a unique character. Many rise steeply from sea level to form impressive buttresses that dominate the skyline, and are visible for considerable distances (Fig. 2.6). Torridonian rocks rest on the Lewisian basement above an unconformity that marks a time gap of about 400 million years.

The name Torridonian was given to these rocks by James Nicol in 1866, originally to describe the sedimentary rocks around Loch Torridon lying above the Lewisian gneiss, although Dr John MacCulloch had identified these red sandstones as a coherent geological unit some 60 years earlier. However, he misinterpreted them as part of the Old Red Sandstone (much younger: Devonian in age). The total thickness is presently about 12km, and it has been estimated that another 3–5km has been removed by erosion, making the original thickness about 16km.

Table 2.2: Torridonian rocks of Northwest Scotland.

Group	Formation	Sediment type
Torridon Group 7000m thick 950 million years old at base	Cailleach Head	Grey shale and red sandstone
	Aultbea	Fine red sandstone
	Applecross	Coarse pebbly red sandstone
	Diabaig (not on Skye)	Breccia and sandstone
Sleat Group 3500m thick (found mainly on Skye, and never in contact with the Stoer Group; undated; weakly metamorphosed)	(probably conformable with the Torridon Group above)	
	Kinloch	Sandstone and shale
	Beinn na Seamraig	Cross-bedded sandstone
	Loch na Dal	Grey shale
	Rubha Guail	Coarse sandstone and pebbly conglomerate
	(unconformable on Lewisian gneiss below)	
Stoer Group 2000m thick 1200 million years old at base	(unconformable with Torridon Group above)	
	Meall Dearg	Cross-bedded red sandstones
	Bay of Stoer	Red mudstones (lake deposits) and volcanic ash (Stac Fada)
	Clachtoll	Conglomerate of Lewisian pebbles filling ancient valleys
	(unconformable on Lewisian gneiss basement beneath)	

Torridonian rocks are subdivided into three groups, and each group is further subdivided into smaller units known as formations (Table 2.2). Both the groups and the formations are named after places in northwest Scotland where the rock successions were first described at so-called type localities.

The sedimentary rocks that make up the Torridonian were deposited as continental beds derived from the weathering and erosion of basement gneisses like the Lewisian and transported to their present position from a landmass that lay to the west of the Outer Hebrides, that is, from the eastern edge of the Laurentian continent (the present-day Canadian Shield and Greenland). Rivers cut valleys into the basement rocks, which at the time formed outcrops of rounded hills resembling the present-day landscape seen in the Lewisian areas of the Northwest Highlands. Relief on this landscape was up to 600m. Since there were no trees, grasses or any land plants at the time, river run-off was high and loose scree deposits were transported across the land surface and deposited on valley floors or as alluvial fans that spread out as sheet deposits where river valleys opened onto flat plains. Beyond the fans, intermittent lakes formed, in which finer sediment accumulated as shale and mud that show sedimentary structures such as ripple marks and desiccation cracks where the muds dried out. The main valleys that received the sediments formed a series of fault-bounded structures. The western edge of the Torridonian outcrop is marked by the Minch Fault, which runs parallel to the eastern shoreline of the Outer Hebrides. Movement on the faults during sedimentation allowed very thick deposits to accumulate. At the time of deposition, Scotland was 10° south of the Equator, in a climate zone of summer drought and heavy seasonal rainfall, which caused deep tropical weathering of the gneiss. The red colour of the rocks is mainly derived from the feldspar content, feldspar being a common mineral in the Lewisian rocks (the red is due to the presence of small amounts of iron in the feldspar), together with hematite cement. In very humid conditions, feldspar normally weathers rapidly to clay minerals; hence the presence of fresh feldspar in the Torridonian is a result of fast rates of erosion and deposition, and short transport distances. Sandstones with a high feldspar content were previously called arkose, but the preferred term is now feldspathic sandstone.

The Stoer Group
The Stoer Group is found only on the Stoer Peninsula, as a narrow belt on the east side of the Coigeach Fault (which can be traced from Rubha Coigeach to the Stoer Peninsula and the Sound of Handa near Scourie to the north). Exposure is excellent around the coast, and there is a fascinating assemblage of well-preserved sedimentary structures, including massive beds, fine lamination, graded bedding, cross bedding, contortions caused by slumping and de-watering, mudcracks, ripplemarks and other original depositional structures. At the base is a very coarse conglomerate containing boulders and pebbles of the underlying Lewisian gneiss, some of which can be matched with Lewisian outcrops lying close by (literally just a few metres away). The Torridonian is seen to fill hollows in the ancient land surface, and there are localities where the original weathered surface of the underlying Lewisian gneiss is still preserved beneath the Torridonian. Much of the gneiss content at the base of the Stoer Group is angular, and the term sedimentary breccia (Italian: 'broken') is often used in preference to conglomerate, in which the pebbles are well rounded. Fresh feldspar is also present as well as quartz, and sedimentary rocks

with angular fragments and feldspar are described as being immature. The material forming the breccia can be shown to have come from only a few kilometres away at the very most.

Red and brown mudstone, conglomerate and sandstone follow above the breccia, and represent deposition by braided rivers in a floodplain that occupied an ancient valley. The rivers dried up periodically, as shown by mudcracks, and the dry sand was then moved by the wind to form dunes on the valley floors and on the surface of dried-out lakes and ponds.

An important unit in the Stoer Group is a red muddy sandstone known as the Stac Fada member, part of the Bay of Stoer Formation. This contains volcanic ash and glass fragments, representing an airfall tuff. The original source (i.e. a volcano that would have been active then) has not yet been found. Since this bed is so distinctive, it makes a marker horizon in the Stoer Group. A hypothesis was put forward in 2008 that this could represent a meteorite impact event. Later (in 2016), it was suggested that a buried geophysical anomaly at Lairg, 50km to the east, may represent the location of this meteorite impact. This view has since been challenged, and a source to the west of Stoer seems more likely (Amor *et al.*, 2019; Sims & Ernstson, 2019). Other components include thin limestones and black shales, and structures that represent gypsum crystals formed by the evaporation of salt-rich water in lakes on an arid land surface. Limestone mud mounds are also present, formed by algal growths on top of rocks in warm shallow water, and thus are the oldest indications of life in Scotland. Taken together, these sedimentary features indicate that the Stoer Group was deposited when Scotland, on the eastern edge of Laurentia, lay at 10° south of the Equator in a climatic zone of hot dry summers and cool wet winters.

The Sleat Group

Rocks of the Sleat Group are found only in the Sleat Peninsula of southeastern Skye and on the adjacent mainland, where they underlie the Torridon Group. The Sleat and Stoer groups are never seen in contact, and it is assumed that the Stoer rocks are older than the Sleat. The rocks are mainly river-deposited coarse grey sandstones and a few shale beds, with a total thickness of 3500m. Because they have been weakly metamorphosed in the Caledonian thrust belt, the Sleat Group has never been accurately dated, and the exact age of deposition is not known.

The Torridon Group

The youngest of the three groups, the Torridon Group, is the most important in terms of thickness and area covered, and the rocks of this group are
responsible for the high mountains of the west coast around Applecross
and Loch Torridon. In fact, this group makes the largest and most extensive
sedimentary rock unit in the British Isles. They were deposited on an old
eroded land surface of hills and valleys in Lewisian gneiss, with relief up
to 600m. A superb example of this unconformity can be seen at Slioch on
the northeast shore of Loch Maree (Fig. 6.2). On the Stoer Peninsula, the
Torridon Group is unconformable on the Stoer Group, the unconformity
representing a considerable time gap. At Loch Assynt it is possible to see
Torridonian rocks lying directly on top of rotted Lewisian gneiss, indicating that the basement was subjected to weathering at the Earth's surface
immediately prior to the deposition of the Torridonian. Exposures of the
unconformity are most spectacularly seen at Enard Bay, near Achiltibuie.
Bedding in the Torridon Group is almost horizontal across the whole of
the region.

At the base of the group is the Diabaig Formation, made of red breccia
and conglomerate containing Lewisian gneiss pebbles, and filling ancient
valleys. Above are sandstones and shales with mudcracks, indicating deposition in rivers and lakes that dried up periodically. The Applecross and
Aultbea formations make up the bulk of the Torridon Group, and differ
mainly in grain size: the Applecross sandstones are coarser than those in
the Aultbea Formation. These rocks were laid down by braided streams in
a rift valley that may have been subjected to frequent earthquakes, judging
by the contorted bedding and slump structures that may have been trig-
gered by sudden earth movements. The Cailleach Head Formation at the
top of the group has shales with mudcracks and red sandstones with cross
bedding and ripple marks, indicating sedimentation in a shallow lake into
which river deltas grew. Fossils in the Aultbea shales are cryptarchs (tiny
spherical forms of algae, but obscure and not well studied), algae and
bacteria. The Cailleach Head Formation, named after the almost inaccessible peninsula guarding the entrance to Little Loch Broom (due west of
Ullapool), is made up of 15 or more cycles of sandstone and siltstone.
Phosphate concretions found in the beds contain microfossils, the earliest
recorded Precambrian fossil find in Britain (in 1885). The 600m of cyclical
beds represent sedimentation in a lake with varying water level, caused by

climate change or fluctuations in the level of the lake floor. Cross bedding and ripple marks are common in the medium-grained sandstone units.

Cambrian and Ordovician

Above the Torridonian lie shallow-water marine rocks – sandstone and limestone – of the Cambrian and Ordovician periods. Again, the relationship is strongly unconformable, representing another time gap of about 400 million years. Rocks of Cambrian to Ordovician age (525–465 million years old) occur as a 2–20km wide strip to the west of the Moine Thrust, from Durness to Skye, a distance of nearly 250km. The rocks are highly distinctive and form unique landscape features. At the base are 200m of white quartzites, followed by a sequence of carbonate rocks (dolostone and limestone) with a total thickness of 800m, and referred to as the Durness Group (sometimes also called the Durness Limestone, although this term is not strictly correct). They were laid down at the edge of a shallow tropical sea fringed with lagoons and islands, when Northwest Scotland was located 15° south of the Equator. Following the deposition of the Torridonian rocks around 950–900 million years ago, they were tilted 10–20° to the west by earth movements. Erosion removed the top 3–4km before the sea invaded and deposited clean white sands unconformably on top of the Torridonian, and in places directly onto the underlying Lewisian rocks. Since the Torridonian is now horizontal, the entire area must have been tilted back by 10 20° to the east, after the deposition of the Cambrian quartzites and Ordovician limestones. The rock succession was first described by Peach and Horne in their 1907 memoir, and their subdivisions are still in use today, with only slight modifications in the names of the formations, most of which are derived from place names near Durness (Table 2.3).

The Durness Group, which spans the Cambrian–Ordovician boundary (490 million years), is often referred to informally as the Durness Limestone, but in fact true calcite limestones (calcium carbonate) are rare, the majority of the rocks in this group being made of various kinds of dolostone (made of the mineral dolomite, calcium–magnesium carbonate). All the rocks in the succession were deposited on the continental shelf, in an intertidal to shallow marine zone. Sedimentary structures include cross bedding and ripple marks that indicate deposition at the edge of a shallow sea. Shelly fossils in these rocks represent the first appearance on Earth of animals with hard parts that could be preserved.

Table 2.3: Subdivisions of the Durness succession (Cambrian to Ordovician age).

Period (age, million years)	Group (thickness, metres)	Formation (and Members)
470 Lower Ordovician	Upper Durness Group 600m	Durine
		Croisaphuill
		Balnakeil
		Sangomore
490		Sailmhor
Upper Cambrian 500	Lower Durness Group 200m	Eilean Dubh
Middle Cambrian 510		Grudaidh
Lower Cambrian	Ardvreck Group 200m	An t-Sròn (Salterella Grit above Fucoid Beds)
520		Eriboll (Pipe Rock above Basal Quartzite)

The oldest rocks, the quartzites at the base, contain the famous Pipe Rock, with numerous vertical worm tubes and burrows, either cylindrical or funnel-shaped and tapering downwards, at right angles to the bedding. These are known as trace fossils, since the soft-body worms that occupied the burrows are not preserved. Cylindrical pipes have been given the name *Skolithos*, and the larger funnel-shaped burrows are known as *Monocraterion*, but in fact the two types may have been made by the same animal. Many of the mountains in the far northwest, such as Foinaven (Fig. 2.3), have sharp summit ridges, clothed with aprons of white scree from these quartzites. Cross bedding is particularly prominent in the oldest of the quartzites, which on some older maps are referred to as False-Bedded Quartzite (false bedding is an erroneous term for cross bedding). Another name that is frequently encountered is Basal Quartzite, while the term Cambrian Quartzite refers to all the Eriboll Formation. During the transformation from loose sand to quartzite, the sediment grains were welded together by a process known as pressure solution, in which a silica cement bonded each sand grain to its neighbours, so that all the intervening pore spaces were filled and the resulting rock is very tough and resistant to erosion. The prominent and extensive quartzite screes on many mountains in the Northwest Highlands contain sharp angular fragments, formed by continuous freeze–thaw action over many centuries on hill tops, including periods when summits were exposed above the glaciers during the last ice age. When broken, quartzite fractures across the quartz grains, as opposed to sandstone, where the cement breaks and the quartz grains fall out intact.

The lack of pore spaces means that drainage is poor, and the rock produces acid soils of low fertility, in marked contrast to the overlying limestone and dolomite beds. Hence, there are very distinctive and often unique plant assemblages on the different rock types in the Northwest, particularly when latitude and altitude are taken into account.

Above the quartzites there is a distinct change in rock type, marked by the Fucoid Beds, a 20m-thick sequence of brown dolomitic siltstones and fine-grained mudstones, showing ripple marks and cross bedding. The rocks are high in potassium and contain beds of iron ore in the form of pea-sized pellets. During the 1980s, a survey was carried out on the feasibility of using the Fucoid Beds as a slow-release potash fertilizer. The name derives from fucoid, meaning seaweed, but in fact the characteristic trace fossils on the bedding planes were made by worms and not seaweed. Other fossils include the trilobite *Olenellus*, allowing the rocks to be dated as late Lower Cambrian, 520 million years old. Trilobites of this age in Wales and the Welsh Borders belong to quite different species, a fact that provided evidence that Scotland was located on the opposite side of an ocean from England and Wales during the Cambrian and Ordovician periods (the Iapetus Ocean, 1000–500km wide), thus the two halves of the British Isles previously formed parts of separate continents.

The next member in the sequence is the Salterella Grit, coarse sandstone and limestone containing abundant remains of the tiny conical shells of the fossil *Salterella*, an extinct group of marine animals. Originally, this fossil was identified as *Serpulites*, and older maps refer to the beds as the Serpulite Grit. Very good exposures of the quartzites, Fucoid Beds and Salterella Grit are found on the north shores of Loch Assynt, from Skiag Bridge eastwards, while the overlying carbonate beds are seen to great advantage at the Stronchrubie Cliffs, Inchnadamph, as well as in the type area around Durness, especially at Smoo Cave.

The Durness Group contains a variety of carbonate rocks (dolostones and limestones), with some chert beds, laid down in shallow lagoons and warm, clear tropical seas at the edge of the continental shelf. They indicate that marine life was varied and abundant. Chert is a form of quartz that originates from the dissolution and redeposition of silica, of organic or inorganic origin, in the form of hard grey nodules and irregular layers during the chemical alteration of dolomite and limestone. Many of the nodules have quite unusual shapes (see Chapter 4).

After their original deposition, the Durness carbonate rocks underwent a complex series of changes during diagenesis, involving recrystallization of the limestone, replacement of calcite by dolomite, replacement of dolomite by silica then by calcite and further development of dolomite. These changes occurred within the rocks by a combination of bacterial action and recrystallization. During these processes, the volume of the rocks changed several times, so that internal fracturing occurred, producing breccias and networks of fine calcite veins. Some of the larger cavities were filled by a variety of calcite crystals known as dog-tooth spar – this can be seen at Balnakeil, near Durness (Chapter 3, locality 6).

The limestones have been responsible for forming a unique aspect of the landscape around Durness and Assynt. Since calcium carbonate dissolves easily in rainwater, streams sometimes disappear down swallow holes, and cave systems form underground. Large expanses of limestone form bare rocky pavements that are eroded at the surface, and joints become weathered out to create fissures called grykes (also grike; Yorkshire dialect: crack or slit in rock), usually harbouring a variety of lime-loving plants. Upstanding square blocks between the grykes are called clints (Norse: klint, part of a crag standing out between crevices or fissures). The general term given to this type of topography is karst, from Slovenia and other parts of the Dalmatian coast.

Moine

Moine rocks (the Moine Supergroup) occupy most of the Northern Highlands between the Moine Thrust and the Great Glen Fault. The term was introduced by Peach and Horne in the 1880s, and took the name of Moine House where they stayed when doing fieldwork in north Sutherland, halfway along the great peat bog of A'Mhòine (Fig. 2.7). They are a thick succession of sandstones, shales and mudstones, metamorphosed to quartzite and mica schist. In a general sense, they dip gently eastwards except where intensely folded, and the Moine block as a whole was transported physically to the northwest by an estimated 100–150km along the Moine Thrust, 425 million years ago in the closing stages of the Caledonian orogeny. The rocks were originally carried as sands and muds by rivers flowing off the Grenville mountain belt to the west (i.e. Canada), and deposited in a shallow sea at the eastern edge of the Laurentian continent, 1000 to 870 million years ago. This sea formed a foreland sedimentary basin, parallel to the mountain

Figure 2.7 Moine House, north Sutherland, in the shadow of Ben Loyal.

Figure 2.8 Vertical Moine schist, showing cross bedding, and isoclinal folds at left of picture.

chain that was being actively eroded. Sedimentary structures such as cross bedding, graded bedding, slumping, mud cracks and ripple marks are abundant in places where the folding and shearing are not too intense (in so-called low strain zones; Fig. 2.8). These features are important in telling the correct way-up of the rocks, and thus can be used to interpret complex sequences of folding, as well as indicating the environment of deposition. To date, fossils have not been found in Moine rocks.

The basement beneath the Moine is Lewisian gneiss, outcrops of which are found extensively within the Northern Highlands, particularly in north Sutherland and in the West Highlands close to the Moine Thrust. Such areas of older rocks surrounded by younger are known as inliers. Where the contact is visible, it is evident that the original rocks that make up the Moine succession were laid down unconformably on this Lewisian basement, which is interpreted to underlie all the Moine rocks east of the Moine Thrust as far as the Great Glen Fault. In most instances, though, the contact is tectonic, i.e. Lewisian and Moine rocks are sheared together. This is particularly the case near Bettyhill in north Sutherland, where it can be an almost impossible task to distinguish the two groups.

In terms of stratigraphy, the Moine Supergroup has been subdivided into three groups, based on rock successions in the West Highlands, around Knoydart and Morar. From west to east and oldest to youngest these are the Morar, Glenfinnan and Loch Eil groups (Table 2.4).

The Morar Group preserves many original sedimentary features, as a result of relatively low deformation, especially to the west of the Knoydart Thrust. Inliers of Lewisian basement rocks are common. A major fault, the Sgùrr Beag Thrust, separates the Morar and Glenfinnan groups on the mainland. Between the Knoydart and Sgùrr Beag thrusts, Morar Group rocks were deformed and metamorphosed at higher grades than those to the west of the Knoydart Thrust. The Glenfinnan Group consists mainly of mica schist (or pelite) and quartzite (or psammite). Deformation and

Table 2.4: Subdivisions of the Moine rocks of Inverness-shire and Wester Ross.

Group	Formation	Rock types
West Highland Granite Gneiss, 870 million years old, intrudes Moine rocks		
Loch Eil Group 5km thick 900 million years old	Loch Eil Psammite	Mainly quartzite, uniform and monotonous
Glenfinnan Group 3km thick 950 million years old	Druim na Saille Pelite	Mainly mica schists
	Beinn an Tuim Schist	
	Lochailort Pelite	
Sgùrr Beag Thrust – tectonic break		
Morar Group 5km thick 980 million years old (oldest)	Upper Morar Psammite	Quartzite
	Morar Pelite	Mica schist
	Knoydart Thrust – tectonic break	
	Lower Morar Psammite	Quartzite
	Basal Pelite	Mica schist
Moine Thrust – tectonic break		
Lewisian gneiss basement (sliced up by thrusts)		

metamorphism have destroyed any original sedimentary structures that may have existed in these rocks. Garnet amphibolites are common, and represent basic igneous dykes, sills and sheets that were intruded before folding and metamorphism. Near the Sgùrr Beag Thrust, the Glenfinnan Group rocks are in contact with Lewisian gneiss basement. To the east, the Glenfinnan Group passes upwards into the Loch Eil Group – quartzites, originally deposited as sands on the beach of a shallow sea, as can be judged from the abundant cross bedding. Current bedding indicates that the source of the sediment lay to the south.

Igneous rocks intruded the Moine schists and quartzites at 870 million years ago. These form two main units: the West Highland granite gneiss, close to the boundary between the Glenfinnan and Loch Eil groups, and a suite of amphibolites, representing metamorphosed gabbros. The West Highland granite gneiss occurs as bodies up to 5km across, scattered across the Moine outcrop for more than 80km, representing a set of granite intrusions that may have formed from the melting of Moine schists before being converted to very coarse banded quartz–feldspar–garnet–biotite gneisses. These gneisses can be seen at Fort Augustus and Strontian, for example, and on geological maps they are named individually after the localities where they occur (e.g. the Ardgour granite gneiss). A fundamentally important aspect of this unit is that it places a limit of 870 million years on the age of deposition of the rocks that form the Moine Supergroup. The amphibolites are quite common, and one of the best-known examples is the Ben Hope sill in northwest Sutherland, which is 200m thick and can be followed for at least 15km. It forms a prominent landscape feature on the slopes of Ben Hope (see Chapter 4).

Since age dates for Moine and Torridonian rocks are similar, it has long been surmised that at least part of the Moine succession represents the offshore and metamorphosed equivalent of Torridonian continental deposits. This correlation was first proposed by Sir Archibald Geikie at the end of the nineteenth century, and supported by Peach and Horne. Detailed studies of the rock chemistry indicate that the upper part of the Torridonian was probably laid down at the same time as the lower parts of the Moine, but in distinct and possibly widely separated basins that were brought together by folding, thrusting and faulting during the early Caledonian events, 480 million years ago. The debate continues, and the current situation is that the Morar Group and Torridon Group can be correlated across the Moine Thrust, and they may have formed parts of the same large sedimentary basin that lay parallel to the edge of the long-gone Grenville mountain chain.

In summary, the extensive tract of Moine rocks in the Northern Highlands records a long and complex history of folding, metamorphism and thrusting that began with the Morarian events 850–800 million years ago, with high-grade metamorphism and partial melting that produced migmatites and granite pegmatites. Pegmatite dykes were intruded at 730 million years ago. During the early Caledonian (Grampian) events, further folding and metamorphism occurred 470–450 million years ago, and finally the Moine rocks were transported west by at least 100–150km in a series of thrust faults, including the Moine Thrust, ending 425 million years ago (late Caledonian or Scandian events). There are still many unsolved problems in Moine geology, relating to age, structure and metamorphism, as well as the nature of the junction with the Lewisian. Research into the age of Moine rocks in north Sutherland, published in 2018 (by Bird *et al.*) suggests that the Morar and Glenfinnan groups are separated by an unconformity, since a metamorphic event at 950 million years ago affected the Morar Group only. This would then suggest that the term 'Supergroup' may no longer be appropriate, since that implies sedimentary and structural continuity.

Dalradian

Dalradian rocks make up much of the Grampian Highlands, between the Great Glen and Highland Boundary faults. They total over 25km in thickness, and are subdivided into the Grampian (oldest), Appin, Argyll, Southern Highland and Trossachs (youngest) groups (Table 2.5). Most of the rocks are metamorphosed sediments, originally conglomerate, sandstone, siltstone, shale, mudstone and limestone; now grit, quartzite, slate, schist and marble respectively (Fig. 2.9). Greenstone horizons represent lavas and volcanic ash from surface eruptions. Pillow lavas at Tayvallich in Argyllshire have been dated at 600 million years – these represent underwater volcanic eruptions. Boulder beds in the Argyll Group (the Port Askaig Tillite) were formed 720 million years ago, during one of a number of Precambrian glaciations.

Ages of Dalradian rocks range from 750 to 400 million years, the youngest being Ordovician. Dalradian rocks are seen around Fort William, Glencoe and Oban and on the journey to or from Glasgow via Inverary,

Table 2.5: Subdivisions of the Dalradian in the western Central Highlands of Scotland.

Group	Subgroup	Beds and rock types (Formations)
Trossachs		Highland Border Complex (ocean floor material)
Southern Highland		Ben Ledi Grit
		Aberfoyle Slate
Argyll	Tayvallich	Tayvallich Volcanics (600 million year old lavas)
		Loch Tay Limestone
	Crinan	Crinan Grit
	Easdale	Easdale Slate
		Scarba Conglomerate
	Islay	Islay Quartzite (= Jura Quartzite)
		Bonahaven Dolomite
		Port Askaig Tillite (glacial, 720 million years old)
Appin	Blair Atholl	Islay Limestone
		Mullach Dubh Phyllite
		Lismore Limestone
		Cuil Bay Slate
	Ballachulish	Appin Limestone
		Appin Quartzite
		Ballachulish Slate
		Ballachulish Limestone
	Lochaber	Leven Schist
		Glencoe Quartzite
Grampian	Glen Spean	Eilde Flags
	Corrieyairack	Various types of quartzite (psammite)
	Glenshirra	(750 million years old)
Dava and Glen Banchor succession	Central Highland migmatite complex	840–800 million years old, highly deformed and partially molten; unconformable beneath the Grampian Group Dalradian; formerly thought to be Moine rocks, probably basement to Dalradian

Figure 2.9 Folds in Dalradian rocks, Kilmory, Argyllshire: folded Dalradian schist, with beds (coloured stripes left to right) cut by vertical cleavage; cleavage fans around folds, which have different shapes in harder and softer beds.

Crianlarich and Loch Lomond (Argyll and Southern Highland groups; Chapter 10). Although the total thickness for the Dalradian is over 25km, a complete succession is not seen at any one place, and it is probable that the rocks were deposited in a number of separate basins that were compressed together by folding, thrusting and faulting during the Caledonian orogeny. Such a great thickness and long timescale (more than 350 million years of continuous deposition) are quite exceptional – improbable even – and the Dalradian may in fact consist of a number of distinct basins filled with sediment and welded together during various events. Closely similar rocks are found in the entire 5000km-long Caledonian–Appalachian belt that stretches from Svalbard (Spitsbergen) in the north to Texas in the south, and there were probably several independent sedimentary basins existing simultaneously along the length of the belt. Dalradian and Moine rocks are never seen in contact, but the oldest groups in the Dalradian are similar in age to some of the Moine rocks north of the Great Glen. From the brief descriptions above, it is obvious that the Dalradian represents a much more varied sequence of sedimentary rocks than the older Moine. For a complete modern description of the Dalradian rocks, see the GCR volume, edited by Stephenson and others (2013).

Caledonian mountain-building events

By 470 million years ago, sedimentation on the margins of Laurentia, including Northwest Scotland, stopped when Laurentia, Baltica and Avalonia collided to form the Caledonian mountain chain. During this collision, the Iapetus Ocean that had separated the continents shrank and closed by subduction. Collision led to folding, thickening of the crust, high-grade metamorphism of the Moine and Dalradian rocks, partial melting at the base of the crust to form granites that were intruded at higher levels, and finally thrusting, left-lateral slip faulting and eventual uplift of the mountain chain, which was then exposed to rapid erosion.

The boundary between the Lewisian and Torridonian on the foreland and the folded rocks within the Caledonian chain is marked by the Moine Thrust zone. The other major faults in Scotland, the Great Glen, Highland Boundary and Southern Upland faults are also Caledonian in origin. They started as left-lateral strike-slip faults, so that the Southern Highlands originated as a series of blocks of crust that were slid into place along these faults. Moine rocks were deformed in two large fold structures, referred to as nappes: the Moine nappe above the Moine Thrust, and the Sgùrr Beag nappe above the Sgùrr Beag Thrust. The Sgùrr Beag Thrust connects with the Naver Thrust in north Sutherland. These two great nappes were pushed westwards across the basement. Each thrust was responsible for movement in excess of 100km, and contributed to crustal thickening by piling up of the nappes, rather like pushing a tablecloth across the surface of a polished table to form a concertina of folds. Along the thrusts, intense folding, crushing and recrystallization led to the formation of mylonite (Greek: milled), a word coined by Charles Lapworth, who introduced the term to geology in 1885 during his work at Loch Eriboll. Alignment of minerals that grew during this event indicates a transport direction to the northwest. Slices of Lewisian basement beneath the Moine were sheared off and became intricately folded with the Moine rocks, especially near thrust contacts. The final movements on the thrusts have been dated at 425 million years, which is effectively the date for the end of the main Caledonian mountain building events in northern Scotland. The Moine Thrust zone marks the edge of the folded mountain chain. Caledonian effects on the Lewisian, Torridonian and Cambrian–Ordovician rocks of the basement were localized and very limited in extent, except where these rocks were caught up in the Moine Thrust zone itself.

In the Northern Highlands, the younger Caledonian events in the period 435 to 420 million years ago are referred to as Scandian (after Scandinavia), whereas the earlier phase of the Caledonian orogeny, at 470 to 460 million years ago, is known as the Grampian (after the Grampian Highlands), during which there was strong deformation and metamorphism at high pressures and temperatures. The last events in the Caledonian orogeny were related to thrusting and the formation of granites that were intruded into the crust in the period 425 to 405 million years ago. These are known as the younger granites, to distinguish them from older intrusions in the Northern Highlands that were affected by Caledonian deformation. In contrast, the younger granites are completely undeformed. In a global sense, the Grampian phase was related to the closure of the Iapetus ocean and the collision between Laurentia and an island arc, while the later Scandian phase occurred when Avalonia collided with Baltica and Laurentia, thereby forming the supercontinent of Laurussia (North America, Europe and Russia) by the middle of the Silurian period, at 425 million years ago.

As more age-dating research is carried out, the number and frequency of metamorphic events is increasing, and it now seems that the Caledonian orogeny progressed in a series of pulses over a long period, rather than the simpler two-fold division into Scandian and Grampian.

Moine Thrust zone

The Moine Thrust zone is a highly complex belt of deformed rocks that marks the boundary between the stable continental basement (the unmoved foreland) and the metamorphosed sedimentary rocks of the Caledonian fold belt (also known as the cover). The Moine Thrust zone extends from Loch Eriboll in the north to Skye in the south in a fairly straight and narrow zone, except for an abrupt easterly indentation in the Assynt district (Fig. 2.1). Erosion has exposed deep levels in the thrust zone in this Assynt 'window', and it is here that considerable progress was made in unravelling the three-dimensional complexities of thrust geometry for the first time. Within the zone are a number of low-angle parallel thrusts, each responsible for transporting a fold nappe westwards. From the base upwards and from west to east in Assynt these were named by Peach and Horne as the Sole, Ben More, Glencoul and Moine thrusts, with the Moine nappe at the top, containing the Moine schists. All of these are wonderfully exposed between Loch Assynt and Loch Glencoul (Fig. 2.10). The zone of complication between the basement and the Moine schists has numerous folds and small-scale reverse faults that effectively sliced up the intervening rocks and moved them in such a fashion that the rock sequence is repeated many times over (up to 20 times in some places). The rock slices were interleaved and stacked one above the other in a similar way to shuffling a pack of playing cards. This particular geometry was first noted by Peach and Horne, and represents a major contribution to structural geology, important on a world scale. Overall, the complex fold, fault, thrust and stacking geometry is referred to as imbrication (Latin: overlapping roof tiles). In more modern interpretations, the whole structure is known as a duplex (Latin: two-fold), containing wedges rotated upwards along reverse faults and bounded by a floor thrust at the base and a related roof thrust at the top. A good example can be seen in Assynt on the Stronchrubie cliffs at Inchnadamph. The geometry implies that all the thrusts within the zone are interconnected and are part of the same system.

Figure 2.10 Moine Thrust zone at Loch Assynt; Sole Thrust runs along shore of loch; Quinag in background, made of Cambrian quartzite (pale grey scree on left) above red Torridonian rocks (right).

Movement along the Moine Thrust was mainly concentrated in the Fucoid Beds of the Cambrian succession, which represented a zone of weakness, allowing the Moine nappe to glide up and over to the west. The total amount of movement on the Moine Thrust is difficult to estimate, but given that there is a thick zone of mylonite, it is likely to be many kilometres, possibly 100–150km displacement to the northwest, over the foreland.

Caledonian igneous intrusions

Within the Caledonian fold belt, and including the Assynt district, Cambrian and earlier rocks are cut by igneous bodies that have rather unusual compositions. These are the Loch Loyal syenite in north Sutherland and the Loch Borrolan and Loch Ailsh intrusions in Assynt. The rock types are described as being alkaline (i.e. with more potassium than is required to form feldspar) and undersaturated (i.e. containing minerals that do not form when free silica is present). In addition to these large bodies, there are many smaller dykes and sills of the same composition. Because of their peculiar mineral contents, the rocks were originally given local names, such as borolanite, ledmoreite, cromaltite and assyntite; other names are based on Norwegian localities, such as grorudite and nordmarkite, and still others from French occurrences, like vogesite. The dominant types are coarse-grained syenite, fine-grained felsite and lamprophyre. Several of the types contain large feldspar or hornblende crystals set in a fine groundmass; these are said to have a porphyritic texture. A famous example is the Canisp porphyry, an intrusive sill found on the lower slopes of the hill of that name in Assynt. The summit of Canisp (847m) is made of Cambrian Quartzite, which rests unconformably on top of Torridonian rocks. Syenite is a plutonic igneous rock (intruded at deep levels in the crust), coarse-grained, containing alkali feldspar (orthoclase) and small amounts of plagioclase feldspar and horn-blende; quartz is absent, or makes up only a tiny percentage of the rock. As the quartz content increases, syenite grades into granite.

It is possible that the igneous intrusions in Assynt may have played a role in the conspicuous bulge in the Moine Thrust zone; this was subsequently eroded to expose the rocks beneath, in the Assynt window.

Igneous intrusion and thrust movements overlapped somewhat in time, and the age for the igneous bodies is 430–425 million years, the same as the last movements on the Moine Thrust. Hence the igneous rocks of the Assynt district were very important in allowing thrust movements to be dated for the first time. The Loch Borrolan and Loch Ailsh intrusions are sheet-like bodies with a steep dip and multiple phases of intrusion, mainly syenite but with some early pyroxene-rich ultrabasic members. Intrusion into the Durness Limestone has altered it to marble (previously quarried at Ledmore), and some of the limestone was incorporated into the igneous rock, thereby forming rocks with highly unusual minerals. The Loch Borrolan complex was investigated in the mid-1980s as a potential source of phosphate, based on its apatite content (calcium phosphate), but no extraction was ever carried out.

The Loch Loyal syenite, the largest (16km²) of only three syenite intrusions in the UK, lies 50km to the north of the Assynt bodies, and intruded the Moine schist country rocks 425 million years ago. Differential weathering of the syenite has produced a majestic mountain at Ben Loyal, which is known as the Queen of Scottish mountains. These alkaline intrusions occur parallel to and just east of the Moine Thrust, and their presence points to an unusual composition of the upper layer of the Earth's mantle beneath this part of Scotland.

Younger Caledonian granites around 400 million years old occur in Argyllshire, and many of these appear to have been intruded where pre-existing lines of weakness are intersected by faults related to the Great Glen Fault. The Rannoch, Etive and Glencoe granites are typical examples.

Devonian and Old Red Sandstone

Overlying the basement rocks of the Highlands are Devonian lavas and sedimentary rocks. The continental sedimentary rocks of Devonian age are referred to as the Old Red Sandstone, and they were deposited by rivers flowing off the Caledonian mountains into low-lying valleys in a semi-arid environment. They occur extensively in Orkney and Caithness and around the southern shores of the Moray Firth, and patchily along the Great Glen and in Aberdeenshire. However, the greatest extent is in the Midland Valley, from Stonehaven to Helensburgh and along the Firth of Clyde. The 400 million-year-old Lorn lavas make up the hilly plateau south of Oban. Small patches are preserved as outliers in the Glencoe and Ben Nevis granite complexes. In Lorn, the lavas are 800m thick and occupy an area of 300km^2, which is much less than their likely former extent. They are mostly basalts and basaltic andesites (higher in silica than basalt) and a few rhyolites (very acidic, i.e. silica rich), erupted from fissures and vents.

Post-Caledonian history

Once the mountain chain had formed, it was rapidly eroded from the Devonian Period onwards, and remained high ground above sea level practically until the Jurassic, when an arm of the sea broke through into the Minches and deposited sand, shale, mud and limestone. Thin beds are found in narrow coastal strips on the east and west coasts. In addition, there are small patches of New Red Sandstone (Permian and Triassic red sandstones and conglomerates) around the north and west coasts, representing deposits laid down in flash floods on a desert floor.

The next major event in the evolution of Scotland was the formation of the Palaeogene (early Tertiary) volcanic complexes, with their basalt lava flows and central gabbro intrusions and dyke swarms, 60–55 million years ago (Fig. 2.11). This event caused the west coast to be uplifted, and subsequent deep subtropical weathering left the upstanding isolated mountains of Assynt, Torridon and Applecross. Glacial activity then smoothed the landscape by removing the weathered material and sculpting the surface (Fig. 2.12).

The sequence of events that were responsible for the creation of the Western Highlands as we see the region today is summarized in Table 2.6.

Figure 2.11 Palaeocene dolerite dyke cutting Dalradian slate.

Figure 2.12 Glacially eroded U-shaped valley: looking out onto Rannoch Moor from Glencoe.

Table 2.6: Summary timetable of geological events in Northwest Scotland.

Time (millions of years ago)	Major geological event
3300	Formation of early crust (intrusion of igneous rocks, eruption of lavas, deposition of some sediments). Scotland part of Laurentia
2900	Folding and high pressure–high temperature metamorphism – formation of Lewisian gneiss complex deep inside the Earth
1200	Uplift of Lewisian rocks to the surface from 35km depth; erosion and formation of a landscape of bare rocky low hills and valleys
1000	Deposition of sandstones from rivers, deltas and lagoons – formation of Torridonian sandstone on land surface; first fossils (soft parts); roughly simultaneous formation of Moine rocks: marine sands, muds and shales, then metamorphosed to mica schist
700	Break-up of Laurentian supercontinent
550	Deposition of sand, shale, mud and limestone on continental shelf – formation of Cambrian quartzite and Ordovician limestone (Durness), appearance of first fossils with hard shells
470–430	Major earth movements create Caledonian mountains as a result of continental collision (early Grampian and later Scandian events)
430	Closing episode of Caledonian orogeny causes movements on Moine Thrust and intrusion of granitic rocks; uplift of mountain chain
425–405	Late Caledonian explosive igneous activity at Glencoe (granite intrusion), Ben Nevis and Lorn (lava flows)
400–350	Extensive period of erosion of young Caledonian mountains, and formation of continental deposits – Old Red Sandstone (Devonian)
300–250	New Red Sandstone (Permian and Triassic) – pebble beds and sandstones in rivers and dunes; Scotland now north of Equator
200–65	Shallow tropical seas around Scotland – Jurassic and Cretaceous; shales, muds, sands and coral limestones, and Chalk
62–55	Intrusion of volcanic centres in the Inner Hebrides; uplift, deep sub-tropical weathering and erosion with huge amounts of material transported from the land surface; establishment of river system (flowing mainly eastwards)
2	Scotland covered by an ice-sheet; further extensive erosion
10 000 years to present	Melting of ice, deposition of glacial sand and gravel; river and coastal erosion to form present-day landscape; glacial rebound causes raised beaches around west coast, and hanging valleys in upland areas

Chapter 3

Tongue to Lochinver

Introduction

The journey north begins at Inverness and continues to Bonar Bridge then Lairg (A9, B9176, A836; 100km). This route, over the Struie Hill (B9176 left at Alness), crosses high ground in Old Red Sandstone, with several thrust faults related to movements on the nearby Great Glen Fault. The thrust slices produce step-like features on the hillside above the viewpoint overlooking the Dornoch Firth. Ben More Assynt can also be seen from here, lying 50km to the northwest. Glacial outwash material (sand and gravel, previously extracted) fills Strath Rory, and on the summit of the moor there are the remains of commercial peat extraction machinery at Aultnamain Inn. The steep and narrow twisting part of the road downhill beyond the viewpoint is a glacial overflow channel, cutting down below the Old Red Sandstone into Moine rocks. The hill above Bonar Bridge (to the east) is the Migdale granite, a late Caledonian intrusion (400 million years old). Migdale Rock was shaped by moving ice, to create a roche moutonnée.

From Bonar Bridge, the road follows the Kyle of Sutherland to Lairg. Lairg dam power station opened in 1958 and raised the level of the loch by 10m, drowning the former shoreline. From Lairg to Tongue (A836; 60km), the road follows Moine rocks practically all the way, except for a narrow slice of Lewisian rocks south of Strath Vagastie. The dominant mountain to the east is Ben Klibreck (962m), a long ridge of migmatite (a mixed rock, with mica-rich Moine schist intruded by many granite veins). At the foot of Ben Klibreck, the Naver Thrust carries Moine quartzite over a Lewisian gneiss inlier, while near the summit (Meall nan Con), the Swordly Thrust is responsible for thickening the migmatite layers. The valley here shows superb examples of glacial sculpting, with the valley sides forming long narrow fingers. Ben Klibreck has a number of well-formed corries on the east-facing slopes. As Loch Loyal is reached, the road enters the syenite

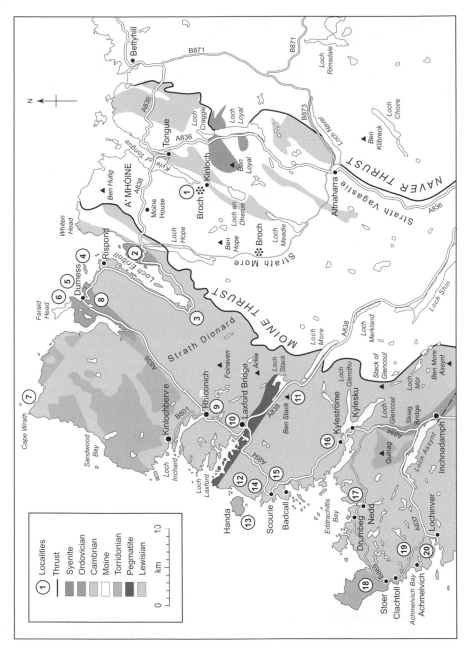

Figure 3.1 Route map with geology: Durness to Lochinver.

Figure 3.2 Ben Loyal syenite intrusion, showing tors at summit.

complex, a large igneous intrusion, and majestic Ben Loyal now dominates the landscape. (Figs 3.1, 3.2)

Continue west towards Tongue for accommodation. From the Tongue youth hostel at the eastern end of the Kyle of Tongue causeway [NC 586 584] there are good views of Ben Loyal (764m) and Ben Hope (927m), as well as the ruined Caisteal Bharraich (Castle Varrich, 14th century Mackay stronghold) perched on a cliff on the headland west of Tongue. West and east of Tongue, coastal exposures of Moine rocks with Lewisian inliers are fully described in the Moine guide (Strachan *et al.*, 2010).

Locality 3.1a Ben Loyal

The Loch Loyal syenite complex is a Caledonian igneous intrusion, dated at 430 million years (the same age as the Loch Borrolan intrusion in Assynt), and the largest area of alkaline rocks (i.e. feldspar-rich, but with little free quartz) in Britain. Smaller alkaline intrusions are found in the Assynt district to the south. The rocks were intruded as a large balloon-shaped body (diapir) into Lewisian and Moine basement gneisses. This forceful intrusion caused deformation and updoming of the surrounding country rocks, which have a fabric superimposed on them. The 16km² intrusion is semicircular in plan, and the tough rocks rise up abruptly from barren moorland (Moine) to form some of the most majestic scenery in the north of Scotland. An Caisteal (The Castle, 764m) is the highest of eight peaks. The 744m peak of Beinn Bheag at [NC 577 483] is also known by climbers as Heddle's Peak, named after Matthew Heddle (1828–1897), an influential Scottish geologist and mineralogist who studied the composition of igneous rocks in the Highlands during the 1880s and discovered many new minerals in the area; his extensive and authoritative collection is now held by the National Museum of Scotland in Edinburgh. Heddle reported the discovery of what he thought was a diamond crystal in mica schist from 5km north-east of Ben Hope, but this turned out to be a colourless variety of garnet. Several quarries have previously worked the white, cream and occasionally pinkish syenite in the past, both for road stone and as building material; the latter can be seen to great effect in the walls of the Tongue Hotel. One such quarry can be visited on the roadside by Loch Loyal at Lettermore [NC 6125 4974], where fresh, pale-coloured medium to coarse syenite may be examined. This is made mostly of feldspar, with small clusters of black minerals – hornblende and pyroxene. Blocks and small fragments of Moine country rocks are found as xenoliths (Greek: foreign rock) in the syenite exposed in the stream just south of Lettermore. The Loch Loyal complex consists of a number of intrusions with varying compositions, at Ben Loyal, Ben Stumanadh and Cnoc nan Cuilean. The earliest intrusions were into a fold in the basement rocks, while the slightly later Cnoc nan Cuilean body was injected along a series of faults. The Loch Loyal intrusion is a special geological conservation site, full details of which are contained in Stephenson *et al.* (2000), while Hughes *et al.* (2013) present a detailed structural map of the complex and a discussion of its origin and chemistry, particularly of the pegmatite veins within the darker syenite at Cnoc nan Cuilean, which contain a number of rare minerals, including those bearing thorium and uranium.

To see the scenery of Ben Loyal to best effect, it is worth taking the minor road south of Tongue to Kinloch; the views from Lochan Hakel are outstanding [NC 5656 5332]. Of particular note are the castellated summits of the mountain (Fig. 3.2). These are know as tors, and originated during the Cenozoic when Scotland was experiencing a prolonged interval of warm, humid climatic conditions during which the rock was rotted at the surface and for some considerable depth, along joint planes. Normally such soft material would have been worn away by glaciers, but during the most recent Ice Age, the summit of Ben Loyal lay above the ice line and remnants of the tors were preserved.

Locality 3.1b Kinloch

Take the minor road for 2km to the head of the Kyle of Tongue and Kinloch Broch (Dùn Mhaigh, mostly collapsed, but the main structures are still visible) at the north end of a prominent rocky ridge overlooking the Kyle and in the shadow of Ben Loyal. Park at [NC 5526 5348] and ascend the ridge, noting on the way the very platy nature of the Moine rocks, which here are quartz-rich (psammites) and contain beds of strongly flattened conglomerate in which the pure white quartzite pebbles have been stretched out into thin pencils in perfect parallel alignment. Garnet–mica schists are also present, in which the deep red garnets stand out clearly on the weathered surfaces. Above these Moine psammites at [NC 5523 5300] there are exposures of green Lewisian hornblende–biotite schist. The boundary between the Lewisian and Moine here is the Ben Hope thrust. Some spectacular examples of folding, including refolded folds, can be found on the flat slabs beneath the broch. This thrust is one of a series in north Sutherland to affect the Moine rocks (the Naver and Swordly thrusts being two others), and they are thought to become shallower at depth and to link with the Moine Thrust some 10km beneath the present-day surface. Considerably more detail on these localities is given in the Moine guide (Strachan *et al.*, 2010).

Return to the main A838 road and cross the Kyle of Tongue by the causeway 2km north of Tongue. It is worth stopping briefly by the slipway at the eastern end of the causeway at [NC 5818 5857] to examine the large fresh blocks of local Lewisian and Moine rocks that have been used in the construction – mica schist, quartzite, banded gneiss, granite, pegmatite and migmatite. Note the way in which the rocks have broken neatly into large flat slabs along the foliation surfaces, an indication that they experienced high levels of strain. Strongly developed mineral lineation on the schistosity surfaces of the Moine rocks is also striking here.

Continue west on the A838 road, crossing the extensive flat moorland of A' Mhòine (Gaelic: peat bog). A brief stop may be made at the roofless Moine House [NC 5182 6003], famous as the inn used by Peach and Horne in the 1880s when they were mapping the area for the first time. They took the name of the house to label the Moine schists. Note the view of Ben Loyal from this point (Fig. 2.7).

As Loch Hope is approached, the road begins to twist and descend quite steeply. Notice how the landscape changes rather abruptly, and that there is much more rock exposed in this vicinity. The Moine Thrust is crossed here, but is not seen at the surface. Above Loch Hope, the cliffs are made of Lewisian gneiss which has been transported to the west across Cambrian quartzites by the Arnaboll Thrust, named after Ben Arnaboll [NC 4590 5900], a prominent craggy Lewisian gneiss hill between Loch Hope and Loch Eriboll, where the thrust can be examined.

Park at the large lay-by at [NC 4681 5970], where there is a tall deer fence with a stile. From here there is an excellent view south down Loch Hope to Ben Hope (927m), the prominent notch almost halfway up being the location of the Ben Hope thrust with the Ben Hope amphibolite sill, an intrusion along the Moine–Lewisian boundary. The geology around Ben Arnaboll is highly complex, due to folding and thrusting. The area is important as the location where Charles Lapworth effectively solved the problem of the Highlands in 1882. It was here that he first interpreted the effects of thrust faults, and introduced the term mylonite to describe the crushed, milled and recrystallized rocks within the thrust zone. On Ben Arnaboll, the Lewisian gneiss has been placed above Cambrian Quartzite by the Arnaboll thrust. Just below the thrust plane at [NC 4612 5964], Pipe Rock is seen in its correct position (i.e. right way up), and the pipes have been deformed by movements on the thrust: instead of perfectly circular outlines, they now have an elliptical shape due to flattening. Above the thrust, the Lewisian rocks have been sheared and converted to mylonite. It was a remarkable discovery for Lapworth to find Precambrian basement sitting on top of the Cambrian sedimentary rocks – quite the opposite of what would normally be expected. The Cambrian rocks are repeated many times over, having been stacked up into a pile of slices during thrust movements. This structural relationship is referred to as imbrication (Latin: overlapping roof tiles). For a detailed description, see the excursion by Butler (in Strachan *et al.*, 2010).

Locality 3.2 Loch Eriboll

Continue on the A838 towards Loch Eriboll. Stop in the roadside car park at [NC 4636 6009] and walk to the flat exposures at [NC 4638 6003]. The white rock here is the Cambrian Pipe Rock, containing worm burrows that have been flattened and stretched in the Moine Thrust zone. Undeformed pipes can be seen at Skiag Bridge, Loch Assynt (locality 4.3). Return to the car and continue to the viewpoint lay-by at [NC 4534 6000] overlooking

Figure 3.3 Ard Neackie, Loch Eriboll; Durness Limestone with lime kilns, and tombolo connecting to mainland. Cambrian Quartzite forms the low feature in the background.

the prominent grassy peninsula of Ard Neackie (a tombolo – an island connected to the mainland by a natural causeway of sand, gravel and pebbles), on which old lime kilns, dating from 1870, a disused limestone quarry and jetty for the former Heilam ferry point are obvious (Fig. 3.3). Durness Limestone was formerly extracted here and on the nearby island of Eilean Choraidh, burnt for agricultural lime (to neutralize acid peat soils), and then shipped out. Loch Eriboll is a glacially scoured fjord, 16km long and the deepest sea loch in the United Kingdom. Plans occasionally resurface relating to proposals to site a coastal superquarry here, from which Lewisian gneiss from nearby Beinn Ceannabeinne and limestone from south of Durness would be shipped.

From the viewpoint, the hills on the western side of Loch Eriboll are made of Cambrian Quartzite, conspicuous from here by the bare white exposures on the slopes. The quartzite rests unconformably on Lewisian gneiss, and forms part of the foreland to the Moine Thrust, unaffected by the Caledonian movements. Dipping east towards Loch Eriboll are the quartzite slopes of Meall Meadhonach, Meall nan Cra, Beinn Spionnaidh and Cranstackie, all of these hills lying above the Lewisian basement that can be seen clearly from Strath Dionard (page 62). The eastern shore of Loch Eriboll has a more complex geology, being within the Moine Thrust zone. Much of the lower ground is made of Durness Limestone, as here at Ard Neackie and at Eriboll Farm, where the fertile soils have been cultivated. Just south of the viewpoint, the cliffs above the road are made of steep and overturned quartzite, transported with Lewisian gneiss by the Arnaboll thrust, which reaches the coast 150m south of the Ard Neackie tombolo. To the south, the prominent hill of An Lean-charn illustrates the gentle southeasterly dip of the Moine schists lying above the Moine Thrust plane.

Kempie Bay, from where the An t-Sròn Formation of the Cambrian succession was first described on the headland of the same name (Gaelic: nose or promontory), is important in the history of geology, as it was the key to resolving the Highlands controversy, since it was here that Lapworth was able to demonstrate the significance of thrusting. In 1883 he proved, by careful mapping, that the lower and upper quartzites were repetitions of the same bed, whereas Murchison had earlier (1860) interpreted the upper quartzite as being younger than the one below, since he failed to recognize the structural complexity caused by folding and overthrusting. This locality is also significant as the standard for the Cambrian rocks of the Northwest Highlands. Butler (in Strachan *et al.*, 2010) describes a very detailed excursion to this classic site.

Locality 3.3 Polla

Park in the lay-by on the south side of the road at Polla [NC 3940 5389], beneath the cliffs of Creag na Faolinn. The rocks at the base of the crags are mylonites from quartzite and gneiss in a thrust, followed upwards by regular Lewisian gneiss. Looking south up Strathbeg from here, there are excellent views to Creag a' Charn Chaoruinn, which is made of Moine schist, carried by the Moine Thrust across Lewisian gneiss on Creag Shomhairle, which in turn is thrust across the Pipe Rock on Conamheall on the west side of the valley. The folds in the Pipe Rock are very obvious from here, particularly the large anticline. Beneath these folds is the Sole Thrust, the lowest one in the Moine Thrust zone. At the mouth of this valley, beside the lay-by, there are extensive sheets of sand and gravel that originated from meltwater streams emerging from the snout of a valley glacier that was here during the most recent ice age. Several of the small lochans on the valley floor are kettle holes, formed when large outlying blocks of ice melted where they stood as the glacier retreated. Butler (in Strachan *et al.*, 2010) describes a strenuous full day excursion along Strathbeg, to examine some spectacular fold and thrust structures.

Continue on the A838 road along the west side of Loch Eriboll, noting the extensive bare slopes of quartzite slabs and the sparse, poor vegetation on the thin soils in this area. The hillsides are liberally strewn with large boulders, carried by the ice sheet and deposited randomly on the surface as the ice melted. Some of the many archaeological remains in the area include a souterrain chamber (an ancient underground chamber, excavated along quartzite bedding planes, and possibly used to store food) close to the roadside at Port nan Con. Splendid examples of glacial phenomena can be seen near Rispond, at the northwest end of Loch Eriboll, where the road turns left towards Durness. Here, the Lewisian gneiss has been smoothed and shaped by the ice into streamlined patterns, and in places polished to form a glaciated pavement, and the cliffs and hilltops characteristically have huge round boulders perched on top.

Locality 3.4 Rispond

Park in the car park above the beach, opposite the road leading to Rispond [NC 4437 6534]. The bare hill above the road is Beinn Ceannabeinne (383m), made of bright pink, red and creamy Lewisian gneiss, with many coarse feldspar-rich pegmatites. Investigations were made in the 1980s with a view to quarrying this rock for use in ceramic manufacture and as an abrasive powder, but no extraction was carried out. Cross the road and walk down the path onto the beach to examine typical exposures of light (feldspar, quartz and muscovite mica) and dark (hornblende and biotite mica) banded gneiss, then go to the west end of the beach to the vertical cliff at [NC 4421 6572], just beyond the fossil sea stack projecting from the sand dunes. Here the gneisses are vertical; note how the black amphibolite bodies are pinched at various places to form what looks like a string of sausages. This structure is referred to as boudinage, and is a very common feature of banded gneisses, where the contrasting compositions of the basic (black) and acidic (pink) units respond in different ways to flattening and stretching during deformation (Fig. 3.4). Deformation took place in the Rispond shear zone, at around 1800 million years ago, in the early Laxfordian event. Stretching lineations indicate a dextral oblique sense of movement. Findlay and Bowes (2017) provide a detailed account of the structural history of this locality, and the area south to Strath Dionard.

From this point there is a good view eastwards to the white quartzite cliffs (Cambrian Pipe Rock) of Whiten Head at the mouth of Loch

Figure 3.4 Boudins of amphibolite and pegmatite in Lewisian gneiss, Rispond, near Durness.

Eriboll. Just offshore, the small island of Eilean Hoan is made of Durness Limestone, the bedding planes of the blue-grey rock being easily made out. Return to the car park and drive west to Smoo Cave.

Locality 3.5 Smoo Cave

Park at the visitor centre above the cave, at [NC 4185 6716]; there are public toilets here. Take the marked footpath down to the entrance to Smoo Cave and enter the chamber, noting on the way the 800m long, steep, narrow inlet with the horizontal bedding planes of Durness Limestone in the Sailmhor (below) and Sangomore formations being clearly visible in the cliffs (Figs 3.5, 3.6). The inlet formed by progressive erosion by the sea along a small fault, and continued roof collapse. This is the largest cave of its kind in Britain, at 60m long, 40m across and nearly 20m high. At the back of the cave are high round mounds of limestone formed, like stalagmites, by evaporation of drips from the roof. After rain, the waterfall inside the second chamber is quite impressive. The swallow hole for this is across the road from the car park, where the stream of Allt Smoo disappears into the chimney that leads to the cave. In dry weather the stream has no water at the surface. It is possible to take a small boat for a short distance into the third chamber, along an underground lake. An extensive cave system

Figure 3.5 Smoo Cave, Durness, showing various units of horizontally-bedded Durness Group carbonates.

Figure 3.6 Durness: swallow hole in limestone above Smoo Cave, where surface water disappears underground into a vertical chimney.

exists around Durness, having been formed by surface water dissolving the limestone, then moving downwards along joints and bedding planes to form underground streams. Additional caves in the Durness Limestone, which have been explored and mapped more thoroughly than the Durness system, can be seen in the Assynt district to the south.

Locality 3.6 Sango Bay

From Smoo Cave, drive 1.5km west to Sango Bay, Durness, via the steep minor coastal road at the Sangomore junction [NC 4110 6738]. Park in the tourist information centre car park beside the Oasis tearoom [NC 4055 6766]. There is a display of large blocks of the local rock types at the entrance to the car park, and an exhibition of geology inside the centre, which also has maps and information leaflets. Note the smooth vertical wall of Durness Limestone exposed in the cliffs on the east side of Sango Bay, while to the far west Moine rocks can be seen. This cliff wall marks a normal fault (the Sangobeg Fault, Fig. 3.7), and the rocks on the beach are made of shattered Lewisian gneiss, Oystershell Rock, Cambrian Quartzite and Durness Limestone in an outlier of the Moine Thrust zone, preserved here by having been dropped down against this fault (Fig. 3.8). The quartzite has been converted to mylonite and is thrust over the younger Durness Limestone. Another such outlier can be seen 3–4km to the northwest at

Figure 3.7 Durness: steep face of Sango Fault, Lewisian to left, Durness Limestone forms face.

Figure 3.8 Sango Bay, Durness; seastacks of black Oystershell Rock within mylonites of Moine Thrust Zone.

Faraid Head, where the Moine Thrust is exposed (see Strachan *et al.*, 2010 and Goodenough & Krabbendam, 2011 for details of the area). The first rock exposures on the beach at the foot of the wooden staircase (low tide is needed) are crushed and sheared Lewisian mylonite, followed by black Oystershell Rock. This is a brittle, flaky rock consisting of quartz, white muscovite mica and green chlorite, with a distinctive wavy structure caused by microfolds, and a dark greyish green sheen on the schistosity surfaces (Fig. 3.9). The name was given to the rock on account of the numerous very thin, intricately folded white quartz veins resembling oyster shells. It is a form of mylonite derived from the deformation of Lewisian amphibolite within the thrust zone. The technical name is phyllonite (Greek: coined from phyllite and mylonite, a rock made of mica-rich sheets or leaves formed by grinding down of coarser rock). The Oystershell Rock forms a number of tall jagged black sea stacks jutting out from the sand (Fig. 3.8). At the western end of the beach are examples of sheared and brecciated red quartzite, followed by limestone stained red by hematite (iron oxide); then quartzite is repeated. All the rocks here are within part of the Moine Thrust zone. At the back of the beach just below the grassy cliffs and near the soft pinkish limestone, the Oystershell Rock becomes a finely striped pale green mylonite, showing many tiny folds. This green mylonite grades up rapidly into the coarser black rock seen in the stacks.

Figure 3.9 Sango Bay, close-up of Oystershell Rock, showing crenulations of quartzite among hornblende and black biotite mica, resembling sea shells.

If there is time it is worthwhile visiting Balnakeil Bay to examine a section through all the units (formations) of the Durness carbonates (see Goodenough & Krabbendam, 2011 for details). In particular, the locality at [NC 3788 6881] shows some spectacular white chert nodules in the blue-grey limestone, with fantastic shapes. The mottled rock known as 'leopard stone' can also be seen here. The dark grey and white mottling effect may have been caused either by soft-bodied animals in branching burrow systems, or by the formation of dolomite during diagenesis, when the soft sediment was being converted to rock. Shelly gastropod fossils (white coils up to 1cm in diameter) can sometimes be found on eroded surfaces at the eastern end of the bay. They indicate an age for these rocks of 480 million years, and can be correlated with fossils in Canada. Numerous small thrusts can be seen cutting the limestone at this locality. Note the excellent view across the bay to Faraid Head, with Moine schist thrust over Lewisian gneiss.

Locality 3.7 Cape Wrath

Return to Durness and drive southwest on the A838 for 3km to the junction for Keoldale and the Cape Wrath Hotel at [NC 3850 6565]. Park by the jetty at [NC 3780 6620] and take the passenger ferry across the Kyle of Durness to meet the minibus to Cape Wrath (timetables can be checked in the tourist information centre at Durness). From the high point on the road at Daill [NC 3611 6816] there is a splendid view of the sand spits in the Kyle, and across to Durness Limestone at Balnakeil Bay and Moine schist on Faraid Head. At Daill itself there are exposures of Torridonian sandstone and Cambrian Quartzite, although the road crosses Lewisian gneiss for the most part. To the north of Daill the Torridonian extends for some considerable distance on the southern slopes of Beinn an Duibhe. Access is not advised, as this is a bombing range with live ammunition possibly lying on the hillsides. The rocks belong to the Applecross Group, and are made of red and brown feldspathic sandstone (arkose) with layers of pebbly conglomerate and angular breccia at the base, overlying the Lewisian gneiss on the north of this hill. Stewart (2002) provides a description of the Cape Wrath member of this group. The minibus stops at the road end [NC 2596 7460], near the lighthouse (built in 1828 by Robert Stevenson, grandfather of the famous author). At Cape Wrath, the cliffs of Lewisian gneiss are some 140m high. From the lighthouse there are good views to the vertical faces of folded Lewisian gneiss, in various shades of red, pink and cream, alternating with bands of black amphibolite, and cut by red granite and pegmatite (see Fig. 2.1). The cliffs just to the east, at Clò Mòr, are made of Torridonian sandstone. At 300m, these are the highest sea cliffs in mainland Britain, and they extend for three kilometres along the coast. Return to Keoldale via the minibus and ferry.

Locality 3.8 Kyle of Durness

Follow the A838 south from Durness, noting the limestone exposed along the roadside. A stop may be made at Sarsrum, opposite the Kyle of Durness [NC 3776 6419], to view the stromatolite mounds (formed by sediment trapped by algae in shallow water), chert bands and the Leopard rock in the light blue-grey Durness Limestone (Fig. 3.10; also seen at Balnakeil). There are several old quarries in the vicinity that were used to obtain limestone for making agricultural lime. The bright green grassy fields are very striking. Here, the straight western side of the kyle marks the line of the NE–SW Grudie Fault. By the bend in the road at the bridge over the River Dionard (Drochaid Mhòr, the big bridge), an impressive corrie can be seen on the hillside opposite, excavated in Cambrian Quartzite. Continue on up the straight road to the brow of the hill near Gualin House for a view of the Strath Dionard dome, a broad open fold in glacially

Figure 3.10 Chert nodules in Durness Limestone, Kyle of Durness.

polished Lewisian gneiss and pegmatite, with the quartzite hills above, cloaked in their white scree. The Gualin National Nature reserve is situated from here down Strath Dionard. At the top of the hill, the Grudie Fault is crossed and the brown Torridonian rocks are seen at the roadside, forming flat slabs, in strong distinction to the knobbly gneiss terrain and the sharp white quartzite ridges opposite. Continue on to Rhiconich, noting the many isolated boulders perched on hill ridges, left behind as the ice melted. Fresh road cuttings expose pale pink and white gneiss, red granite, and black amphibolite sheets in the Northern district (Rhiconich terrane) of the Lewisian outcrop. These rocks have been dated at 1850 million years old (Goodenough *et al.*, 2013).

Locality 3.9 Loch na Fiacail

Loch na Fiacail (Gaelic: loch of the tooth); park at [NC 2329 4864] in the lay-by with a SNH information board. Take care when crossing the road at this locality; do not hammer the exposures (for safety and conservation reasons). The grey and cream banded gneiss (2800 million years old) and pink to white migmatite have been folded, then cut by black amphibolite sheets, with a very strong fabric. These have been interpreted as highly altered Scourie dykes, which have been cut in turn by granite (1855 million years old), pegmatite and quartz veins. Boudinage structures (pinch-and-swell) are common at the margins of all the rock types. By looking

Figure 3.11 Loch na Fiacail, north of Loch Laxford. Banded Laxfordian gneiss (grey), cut by amphibolite sheet (black), in turn intruded by pegmatite (pink).

very carefully, some cross-cutting relationships can be seen between the amphibolites and the grey gneisses (Fig. 3.11). Loch na Fiacail marks the northern margin of the extensive suite of Laxford granites: examples of distinct granite dykes with sharp margins decrease northwards, merging into patchy and indistinct migmatites among pink and white quartz–feldspar gneiss. An age of 1750 million years has been obtained for a high-temperature metamorphic event during the Laxfordian episode, which is probably associated with the coming together of the Northern and Central regions (Rhiconich and Assynt terranes) along the Laxford shear zone (Goodenough *et al.*, 2013).

Locality 3.10 Loch Laxford

Park at the lay-by beside the boat house [NC 2278 4781] for a view along the length of the inlet to Loch Laxford. In the high road cutting at the corner are folded migmatite, pink and white gneiss, pegmatite and granite veins (the Laxford granites) intruded along fold axial planes (Fig. 1.1). Some veins seem to appear from or disappear into the banding, indicating partial melting of the gneiss and local generation of granite, at about 700–750°C. Here, the Laxford granite sheets mark the northern edge of the Laxford shear zone (sometimes referred to as the 'Laxford Front'). This is a zone of weakness, where the steep slabs of gneiss have been excavated by the ice to form a fjord.

Locality 3.11 Loch Stack

Continue to Laxford Bridge, turn left onto the A838 to [NC 2760 4315] opposite Loch Stack and beneath Ben Stack, which is made of Lewisian granite and pegmatite sheets, for a view to Arkle, where Cambrian Quartzite is unconformable on top of Lewisian, the line being clearly marked at the base of the white scree (see Frontispiece). The quartzite has been piled up along thrusts that are practically parallel to bedding planes, resulting in a great increase in thickness of the quartzite.

Gannon (2012) describes the geology on a walk to the summit of Ben Stack from here. This hill is made of granite sheets dipping at 70–75° to the southwest, as can be easily seen from Laxford Bridge. Return to Laxford Bridge and turn left onto the A894. Views along the road are to the quartzite mountains north of Loch Laxford (Arkle, Foinaven, Cranstackie) rising above the knock (from Gaelic cnoc, a rocky hill) and lochan topography of the Lewisian. At Badnabay, the road follows the strike of strongly foliated gneiss, pegmatite and granite sheets, all closely parallel and dipping steeply to the southwest. Strong southeast-plunging lineation on gneiss foliation surfaces at Weaver's Bay on the south shore of Loch Laxford (a narrow inlet along a fault line) is very obvious. Park in the large lay-by at [NC 2082 4765], cross the road and climb a short way up the low hill for a spectacular view of Loch Laxford (bare rocky islands, pink and red granite and pegmatite), with Cranstackie in the distance to the north, then Foinaven and Arkle, all topped by quartzite scree, and Ben Stack in the near distance (see Fig. 2.2). This point in the Laxford shear zone more or less marks the limit of the Laxford granite sheets, which disappear quite suddenly south of here.

At Clashfern [NC 2038 4667], in the middle of the Laxford shear zone, there are distinctive brown-weathering schists and gneisses that break into thin slabs. These represent a belt of metasedimentary rocks, rich in quartz, feldspar, biotite, garnet, iron oxides and sulphides, in places garnet and kyanite. They were originally greywacke, shale and sandstone, now intensely flattened and sheared, with a very pronounced schistosity that dips steeply to the southwest. Interleaved between the schists are dark banded gneisses and isolated fragments of coarse, dense dark green metamorphosed basic and ultrabasic igneous rocks in the form of sheets parallel to the banding in the gneisses. These rocks may be the dismembered fragments of layered ultrabasic–basic igneous intrusions. A distinct mineral lineation can easily be seen on the schistosity surfaces, plunging steeply to the southeast. The many Scourie dykes here often contain very large clusters of blood-red garnet crystals. The schists occupy tight synclinal folds, while the dark basic gneiss forms anticlines. In this area, there is a strong association of metasedimentary rocks, ultrabasic igneous rocks and shear zones. The shear zones are cut by dykes, which are themselves folded and sheared; all the rocks are then cut by granites that increase in frequency towards Loch Laxford. For an excellent general review of the geology between Laxford and Scourie, see Wheeler (2010).

Locality 3.12 Tarbet

Drive along the narrow single-track road for 3km to Tarbet, following the strike of the Laxford shear zone, past Loch nam Brac (Gaelic: trout loch). Notice the steep slabs of dark gneiss and schist on the hillsides. Park at Tarbet opposite Loch Dubh, [NC 1631 4886] (toilets, restaurant, Handa ferry). The gabbro intrusion of Cnoc Thaigh Adhamh (hill of Adam's house) forms the prominent black hill above Loch Dubh. On the beach to the right of the jetty the flesh-pink coloured rocks (quartz–feldspar–biotite–garnet schist) contain numerous small folds (low tide needed). More spectacular examples can be seen by taking the footpath around the back of the restaurant and walking northwards. At [NC 1640 4935] there is an excellent example of refolded red granite veins (dated at 2700 million years old) in dark brown gneiss on a vertical wall by the path (Fig. 3.12).

Continuing north on the coastal path towards Rubha Ruadh (Gaelic: red point), there are many examples of narrow shear zones in the metasedimentary quartz–mica–feldspar schists, several phases of dyke intrusion in various directions, differing in composition and thickness, showing several fold phases and minor shears, layered basic and ultrabasic igneous rocks. Granite sheets and pegmatites begin to appear, increasing in frequency and width towards the north. There are a number of steep gullies and high overhanging inland cliffs resulting from weathering of the schists, so great care must be taken on this fairly strenuous route, which should not be attempted in wet and misty weather. Examine the prominent dyke at Rubh' an Tiompain [NC 1606 4965], where there are many horizontal and vertical faces, providing perfect three-dimensional exposure. This dyke has multiple phases of intrusion, and is cut by many very narrow shear zones where the random igneous texture has been converted to a schist fabric,

Figure 3.12 Tarbet, near Scourie: folded gneiss in Laxford Shear Zone.

Figure 3.13 Basic Scourie dyke cutting banded gneiss within Laxford shear zone; Rubh' an Tiompain, north of Scourie.

with a pronounced mineral lineation on the vertical schistosity planes, consisting of needles of hornblende and streaks of creamy feldspar (Figs 3.13, 14A, B). Do not go round the steep vertical cleft, but instead follow the dyke inland, noting how it suddenly thins and pinches out completely. Looking back to the coast, the cross-cutting nature of the dyke margin is quite obvious, although the angle of discordance is not very great. From

Figure 3.14A Detail of dyke in Fig. 3.13: pronounced lineation shown by feldspar and hornblende streaks, and narrow shear zones in dyke.

Figure 3.14B Multiple intrusions of magma in dyke shown in Fig. 3.13.

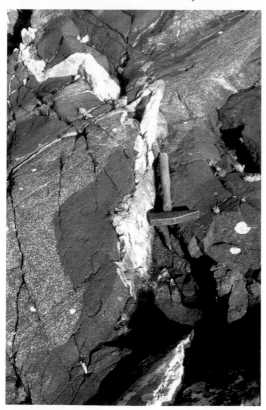

here head to Cnoc Gorm (Gaelic: blue hill) [NC 1680 4974], a metagab-bro intrusion, containing large patches of deep red garnets surrounded by rims of white feldspar. Note how the elongate hornblende crystals form a strong lineation. This rock, which belongs to the early basic suite of intrusions, continues several kilometres southeast to the Moine Thrust (see Fig. 2.4). Good views can be obtained to Foinaven, Arkle and Ben Stack from here. Continue north to a point on a hill [NC 1656 5081] overlooking the broad flat surface on top of the Rubha Ruadh sheet, a 1km-wide coarse-grained red granite dyke containing numerous patches of coarse, bright pink feldspar–quartz–mica pegmatite. From this point northwards, there is a sudden increase in granite, pegmatite and migmatite (mixed acidic and basic gneiss), and Rubha Ruadh is taken as the boundary between the quartz–feldspar gneisses of the Northern region (Rhiconich terrane) and the granulites of the Central region (Assynt terrane). Follow the strike of the rocks to meet the road at Fanagmore [NC 1785 4981] and walk back to Tarbet, passing Loch Gobhloch and Cnoc Thaigh Adhamh on the way. At Loch Gobhloch [NC 1740 4945], note the large patch of soft crumbly bright green, orange, brown and rusty-red schist – these are iron-rich metasedimentary rocks. The hill on the right of the gully (Bealach Tharbait) has exposures of metamorphosed limestone and quartzite among the metasediments (Cnoc na Cuthaige, Gaelic: cuckoo hill).

Locality 3.13 Handa Island

For an optional half day extra tour to see the Torridonian on Handa Island (Scottish Wildlife Trust), take the passenger ferry from Tarbet to the landing spot on a sandy beach. Along the Sound of Handa there is a fault between the Torridonian and Lewisian, dropping down the Torridonian. The walk takes about three hours to cross the island and back, for a view of vertical sea cliffs, natural arches, sea stacks and caves, resulting from weathering of joints and bedding planes (Fig. 3.15). On the way, note the pebble beds with quartzite and pink gneiss clasts. The rocks here belong to the Cape Wrath Member of the Applecross Group. Views back to the mainland show the stark contrast in landforms between Torridonian (gentle topography, heather moor inland) and Lewisian rocks. Handa has one of the largest seabird colonies in northwest Europe. The village was inhabited by 60 people until 1847.

Figure 3.15 Handa Island, opposite Tarbet, Scourie: horizontal cliffs and sea stacks of Torridonian rocks.

Locality 3.14 Scourie

Drive from Tarbet to Clashfern and turn right on the main road to Scourie. Note how the banding now becomes much gentler in dip – the edge of the Laxford shear zone lies just beyond Claisfearn; the number and width of smaller shear zones decrease from here southwards. At the top of Scourie Brae, there is a well-exposed flat-lying boudinaged ultrabasic sheet (black), again possibly part of a dismembered layered intrusion. Continue down Scourie Brae towards Loch a' Bhadaidh Daraich ('oak tree hollow'). The thick layering in the grey granulites is very obvious on the cliffs at the back of the loch.

Drive to the north side of Scourie Bay; park at Scourie harbour [NC 1550 4495], and examine the exposure that forms the wall at the back of the pier. This is garnet peridotite, a tough, very dark khaki-brown weathering, coarse, even-grained, homogeneous, layered igneous rock with 15cm lenses of large blood-red garnet and black pyroxene crystals. Fresh surfaces are jet black and lustrous. The margins of the intrusion are parallel to the banding in the surrounding granulite gneiss. Note the elliptical shape of the garnets, and the fact that they are surrounded by a greenish grey rim (or corona) of feldspar as a result of alteration. At Poll Eòrna, the notch of the Scourie dyke at Creag a' Mhàil can be clearly seen [NC 1501 4557] (low tide required; otherwise the dyke can be examined at the back of the boulder-strewn beach near the stile). This is the type locality of a Scourie dyke, described by Sir Jethro Teall (1849–1924) in a famous paper published in 1885, where he described for the first time the alteration of dolerite to hornblende schist due to stress. The 35m-wide dyke can be picked out by the weathered notch. It is obvious that the dyke cuts almost vertically across the bands in the surrounding granulites, and is therefore younger. At the margins there is fine-grained hornblende schist, but the centre of the dyke has a random igneous texture. Both margins of the dyke are sheared, as is the adjacent country rock, with the fabric curving into the shear zones. Return to Scourie jetty.

For additional excursions to the dykes at Scourie graveyard and the large layered ultrabasic intrusion at Camas nam Buth, Scourie Mòr, see Goodenough and Krabbendam (2011).

Locality 3.15 Badcall Bay, Eddrachillis

From locality 3.14, rejoin the main A894 road, turning right and proceeding south to the junction signposted for Upper Badcall. Turn right on to this narrow, twisting road, past Loch an Daimh Mòr (Gaelic: the big deer). At the brow of the hill, just opposite a house, there is an obvious bright green and silvery-grey cutting by the roadside, on the right [NC 1601 4254]. This is soapstone, a very soft rock (it can be scratched by a finger nail), consisting of talc that originated from the alteration of an olivine-rich ultrabasic rock. Loose material can be collected, but please do not further reduce this small and unique locality by hammering. It is reputed to have been worked in prehistoric times as a quarry for use in making carved ornaments. From here there is also a superb view south across Eddrachillis Bay to the glaciated Lewisian gneiss platform, above which rise the isolated Torridonian sandstone hills that make this landscape unique. The bay itself is scattered with many islands and skerries of gneiss.

Proceed now to the end of the road and park carefully at the back of a large lay-by that also serves as a turning place [NC 1532 4170]. Go through the gate then onto the path towards the shore, bearing left past a number of ruined croft houses and one that is still occupied, at the end of the track, through two more gates. The ground is quite boggy after rain. Near the shoreline, take great care on these rocks, as the surface dips seaward and is slippery in wet weather. Do not walk on these rocks in storms. In this vicinity, the granulites are typically coarse, grey, even-grained, with no oriented fabrics, but there is a roughly expressed compositional layering of lighter and darker rocks. When weathered, the surface is speckled green and white. The rock surfaces are covered in white, light grey and pale orange lichens. Compositionally, the rocks are made of pyroxene, feldspar, quartz and garnet and when fresh they have a glassy lustre.

North of a distinct gully at [NC 1507 4158], the grey gneiss is interbanded with dark hornblende–plagioclase lenses, balls and pods drawn out into elongate elliptical shapes. Near Geodh nan Sgadan, at [NC 1461 4156], a basic dyke with large garnet crystals cuts the gneiss, and has a shear zone at its margin. The cleft itself is an eroded shear zone. Adjacent to this, a pink pegmatite sheet cuts the gneiss, which here is what appears to be a jumbled mass of angular pieces of hornblende rock cut across by white feldspar–quartz patches, sheets, veins and nebulous nets. Such a rock is referred to as agmatite (Greek: rock made of fragments), and probably formed when an original basic igneous rock was injected by granite veins and broken up, then surrounded by an irregular mass of granite. If this agmatite is followed out, it can be seen to be folded, stretched and thinned, to such an extent that the angular nature disappears and the end product is a black and white banded gneiss.

Now walk round the bay towards the southwestern point of this peninsula, to Farhead Point, where there is a very distinct inlet of the sea with an excellent view from [NC 1500 4108] across the small bay to a spectacular shear zone where the horizontal gneiss suddenly turns up into the vertical. Within this shear zone, on the narrow footpath (take care), it is possible to see how a basic dyke and the host gneiss have been thinned and converted into fine-grained, strongly foliated schist. This locality is described in more detail by Beach (in Barber *et al.*, 1978). Return to the car parking space, which is visible from this point, then drive back to the main road and turn right towards Kylesku.

Locality 3.16 Kylesku

Along the roadside, the rocks exposed in cuttings are dark green to black granulite, with a gentle dip, sometimes folded, and with narrow shear zones cutting the banding. Before the road descends down the valley towards the forest, stop briefly at the roadside viewpoint [NC 1928 3777] for a good view of Loch Glencoul with the Glencoul Thrust, the Stack of Glencoul (with Moine schist above the Moine Thrust) and the

Torridonian mountain of Quinag. Continue on for another 3km and stop north of Kylestrome at the Kylesku Bridge viewpoint [NC 2113 3502]. The rock on the roadside opposite is quite typical of the central region of the Lewisian: black to very dark green, poorly banded, homogeneous, coarse, even-grained (random, not oriented) basic granulite, very dense, with a greasy look and feel on fresh surfaces. The mineral content is pyroxene, olivine, feldspar, garnet and iron ore.

If there is time, a half-day boat trip to the Stack of Glencoul is worth the effort. Turn left and follow the road to the Kylesku jetty [NC 2300 3375], where there is parking, a hotel, toilets and telephone. Arrangements must be made with the boatman to land at the head of Loch Glencoul, and be collected later in the day. Do not attempt this trip in bad weather. The headland of Aird da Loch between lochs Glendhu and Glencoul is made of knobbly Lewisian gneiss. There is then a prominent ridge dipping into Loch Glencoul, which is made of white, well-bedded Cambrian quartzites followed upwards by Fucoid Beds and Durness Limestone, lying unconformably on top of the basement gneiss. Above this, on the hill of Beinn Aird da Loch, the knobbly Lewisian appears again. Since this is older than the quartzite and limestone, it must have got there by thrusting along the weak layers of limestone. This is the Glencoul Thrust, and it can be seen just before the boat sails into Loch Beag, in the cliffs at Tom na Toine [NC 2125 2511]. Brown Fucoid Beds and creamy yellow Durness Limestone are very obvious here, immediately beneath the thrust. Vertical orange-brown Fucoid Beds in an imbricate zone (multiple thrusts carrying thin slices one on top of the other) are well seen on the low islands at the head of the loch. If landing, the drop-off point will depend on the state of the tide, as there are two jetties. In any case, take the obvious wide footpath from [NC 2710 3055] up the glen, following the Glencoul River for about 2–2.5km until a safe crossing point can be reached at the foot of the Stack, near [NC 2917 2893]. This will be straightforward unless there has been heavy rain.

Proceed uphill to the Stack and walk around the base to the west-facing cliff. The reward is a stunning exposure of the Moine Thrust – it is knife-sharp, with greyish-green, finely striped Moine mylonite above and Cambrian Pipe Rock below (Fig. 3.16A, B). The pipes in the quartzite have been so strongly deformed that they now appear as white pencil shapes with an extremely flat elliptical cross-section, instead of the original perfect circles, as can be seen at Skiag Bridge (locality 4.3). No hammering

Figure 3.16A Stack of Glencoul: Moine mylonite above Cambrian Pipe Rock mylonite, location of Moine Thrust.

Figure 3.16B Detail of highly strained Pipe Rock below Moine Thrust at the Stack of Glencoul; the pipes are flattened into thin white streaks.

or collecting is allowed, including fallen material. Sadly, almost all the deformed Pipe Rock has been removed by blasting, leaving nothing for others to see in the field. Flattening and thinning has reduced the beds to less than half their original thickness, but the strain decreases rapidly downwards, and only 30cm below the thrust the rocks are practically undeformed. The locality is a world classic site for the study of strain in thrust belts (see Butler in Strachan *et al.*, 2010, for a detailed discussion of the structure and significance of this locality).

From here there is a good view across Loch Beag to the highest waterfall in Britain, Eas a' Chual Aluinn (200m), tumbling over Lewisian gneiss. Return to the pre-arranged pick-up point and back to Kylesku. While in the vicinity, it is well worth visiting the excellent Rock Stop – the Geopark visitor centre at Unapool, with a view towards the Glencoul Thrust, geological displays, toilets and a cafeteria [NC 2374 3275].

Regain the main road, turn left then stop after 2km, by the large lay-by on the left at Newton, where there is a SNH information board [NC 2355 3204]. This is the viewpoint for the Glencoul Thrust (Fig. 1.4), with Lewisian at the top of the hill, thrust over thin limestones above well-bedded pink quartzite, which in turn rests unconformably on top of Lewisian – i.e. the lower boundary is a sedimentary contact. On the skyline, the Stack of Glencoul is very obvious – this is a slice of Moine schist carried over Cambrian Quartzite by the Moine Thrust. Looking south, there is an excellent view of the Torridonian beds towering above on Quinag. On the other side of the road at this point the gently dipping Lewisian gneiss banding is cut by a vertical black basic Scourie dyke. Continue up the hill past Newton and turn right after 750m onto the narrow twisting B869 road, signposted for Drumbeg. This road is steep at certain points with few passing places, and in summer progress can be very slow, with much reversing called for.

Locality 3.17 Drumbeg

Along the Drumbeg road note the flat dark Lewisian granulite, cut by many Scourie dykes. Stop at Drumbeg viewpoint and look north to Eddrachillis Bay, Foinaven, Quinag and Handa island. At the western end of Loch Drumbeg, park by the sheep pens at a bend in the road [NC 1148 3279] beside Achloist, and examine the low, smooth, ice-rounded exposures of black to very dark green rock with obvious horizontal layering (note large patches of white lichen on the surface) on the shore of the loch around [NC 1166 3291]. This is part of a 100m thick igneous intrusion, showing basic gabbro (with pyroxene, garnet and feldspar) and ultrabasic peridotite (with olivine and pyroxene) layers. The layers may have been formed by settling out of heavier crystals under gravity. It is cut by several narrow

Figure 3.17 Drumbeg: layered ultrabasic igneous intrusion; layers are horizontal, and are cut by narrow vertical shear zones that have been altered.

vertical shear zones, in which the granulite has been thinned and converted to hornblende schist. Being less resistant, the shear zones are often weathered out. The sheet was intruded 2900 million years ago near the base of the crust, 45km deep, at a temperature of 1250°C, and was then cut by the materials that make up the grey gneiss and metamorphosed to granulite. It is one of the oldest parts of the Lewisian complex and may possibly represent material from the ocean floor that was attached to the continental crust as it was forming. The original olivine crystals have been altered to a mixture of talc, tremolite and chlorite, all of them soft, light-green minerals that weather out easily, to produce deep hollows (Fig. 3.17). Observe how the large garnet crystals, 5–6cm across, are surrounded by a reaction rim of lighter feldspar. The country rock may be examined on the track to Achloist, where the coarse-grained pale grey granulite is mostly massive, and only vaguely banded. At the corner opposite the lay-by, the country rock is darker (i.e. more basic), with rusty weathering seen along joint planes, and shows banding dipping about 30° to the north-east. This locality is more fully described by Park (in Mendum *et al.*, 2009).

Continue to Stoer, and note that just after Clashnessie the landscape changes abruptly from the knobbly Lewisian terrain to flatter ground, where Torridonian sandstone crops out.

Locality 3.18 Stoer

The Stoer peninsula consists of Torridonian lying unconformably on Lewisian. This is the type area for the Stoer Group (see Table 2.2), and is well described in an excellent memoir with field guide by Stewart (2002, and in Mendum *et al.*, 2009) and in Goodenough and Krabbendam (2011).

At Stoer village, the junction between the Torridonian and Lewisian can be seen at the old cemetery, by a bend in the road just south of the roofless church [NC 0411 2842]. Take the footpath that leads through the cemetery, and go through the gate. At the base, the Torridonian is made of a jumble of angular and rounded boulders, cobbles and pebbles of Lewisian, lying directly on top of the main outcrop of Lewisian. This particular point is important in marking the type locality for the base of the Stoer Group.

Cross the road and go up the hill slightly, behind the white house (Stoer Lodge or Stoer House) and walk across the field towards some low sea cliffs. From the top of the rise there is a clear view to a long finger of rock jutting out into the sea, just beside a shallow bay with white sand and turquoise water beneath a bright red cliff and cave (red limestone, not accessible). This is the rock of Stac Fada (Gaelic: the long rock; Fig. 3.18A, B), the type locality of the Stac Fada member of the Stoer Group (Table 2.2) and of exceptional importance. It is straightforward to walk to this rock, at [NC 0333 2854]. It is a greenish mudstone, with numerous dark, angular volcanic fragments randomly scattered on the bedding planes. These were originally crystals erupted in a volcanic event and carried by the wind before landing on the surface of a lake. Round pea-shaped material at the top of the bed is also volcanic in origin, having formed as lapilli (Latin: small pebbles) in a thunderstorm when dust was falling through rain clouds. No source of the volcanic eruption has ever been found. Much better examples can be seen at Enard Bay (locality 4.12). The lower part of the unit is thought to have formed as a mudflow rushing downhill, as it is composed of a jumbled mass of Lewisian gneiss pebbles and blocks of sandstone. Another interpretation is that volcanic material was injected into wet sediment, as some of the sandstone is

Figure 3.18A Stac Fada, Stoer village. Stoer Group Torridonian rocks; sea cave eroded into limestone beds. The stack is a bed of volcanic ash (B).

Figure 3.18B Shards of broken crystals in Stac Fada member.

distorted. A further suggestion of a meteorite impact will be discussed at Enard Bay, locality 4.12. Please do not hammer or scratch any of the rocks at this locality.

Continue for 1km to Clachtoll (Gaelic A' Chlach Thuill: the rock with the hole) and park in the public car park at the entrance to the caravan site, beside the nature warden's hut; there are toilets here [NC 0394 2729]. A look around the area will show immediately that there are patches of Torridonian lying in hollows amongst knobbly Lewisian outcrops, reflecting the fact that an older landscape is being re-exposed at the surface. The famous Split Rock, which gives Clachtoll its name, is visible on the shore from here.

Walk across the beach to the south to a rocky slope dipping towards the beach at [NC 0411 2709], to see the basal unconformity of angular Lewisian breccia on top of the gneiss bedrock, representing a fossil scree some 1200 million years old, when the Lewisian was exposed at the surface. These are the oldest Torridonian rocks in the country. The large, high red rock projecting from the sand in the middle of the beach is Torridonian mudstone and sandstone. Follow the unconformity around the bay (at low tide) and in the next little sandy bay to the south [NC 0402 2693], the junction between the Lewisian and Torridonian Stoer Group is superbly exposed. Unfortunately, it has been defaced by indiscriminate drilling of sample holes along its length. The gneiss here is very steep, with black balls of ultrabasic rock, and shows many clear examples of folds.

Now walk southwest to the rocky headland at A' Chlach Thuill [NC 0384 2673], where there is perfect exposure of Torridonian sandstone, mudstone and red limestone (Fig. 3.19). On the way, note the lazy beds – furrows and ridges in the machair soil that were used in the past for growing crops. Sedimentary structures here are very abundant, and include cross bedding, mud cracks, ripple marks, graded bedding and

Figure 3.19 A' Chlach Thuill (split rock, Clachtoll): red Torridonian sandstone beds, showing bedding planes and vertical joints.

slumping, all related to deposition of sand into shallow water, probably a lake on a river floodplain that periodically dried up.

Locality 3.19 Roadside viewpoint

Return to the car park and drive southeast across Lewisian rocks, noting a particularly striking low ridge of bright green rock on the left, parallel to the road – this is an ultrabasic dyke that has been altered to chlorite–talc–tremolite schist. From here on the road is steep and narrow, with a very sharp bend at the top, so great care is needed. Stop at the lay-by and viewpoint [NC 0756 2555] on the right hand side to admire the land-scape of Torridonian hills (Canisp, Suilven and Cùl Mòr) rising above the Lewisian basement (see Fig. 2.2). On the roadside opposite, note the steep foliation of the gneiss. Carry on for another kilometre and turn right at [NC 8029 2478] towards Achmelvich. This road is very narrow, with few passing places. After 3km turn right at the telephone kiosk and park near the Youth Hostel and warden's hut in the large car park at the end of the road. There are toilets here [NC 0585 2480].

Locality 3.20 Achmelvich

Exposure on the beach and hills around Achmelvich is very good, allow-ing easy examination of the Lewisian rocks. The locality is important for being the type locality of the Inverian event of folding, shearing and metamorphism that affected the granulites formed in the earlier Badcal-lian event (and previously referred to as the early Scourian; see Tarney, 1963). The 1.5km wide vertical Canisp shear zone cuts through this area. It is the largest shear zone in the Central region of the Lewisian and runs 15km southeast from Achmelvich to Canisp, where it is covered by Torri-donian rocks. It also has a long history of movement (see Hardman, 2019 for an excellent summary). In the middle of this Inverian shear zone is a later narrow belt of shearing that occurred during the Laxfordian episode. Walk across the dunes to the wave-washed and sand-blasted rocks on the north side of the bay, by the dazzling white sands. Here there are good examples of all the typical features of the Central region (Assynt terrane) gneisses, with small-scale folds, and numerous black hornblende–pyrox-ene balls surrounded by the gently dipping foliation of the gneiss (Fig. 3.20). Some of the pods are zoned, khaki-brown to dark green at the centre (altered olivine) and dark blue to black on the outside (pyroxene).

Figure 3.20 Achmelvich: ultrabasic pods and balls (black) in Lewisian gneiss.

Figure 3.21 Achmelvich: folded gneiss at edge of Canisp shear zone.

At the top of this low headland, several narrow vertical shear zones cut the gneiss, and within these, the balls have been stretched into many different shapes. In places the gneiss looks wispy and patchy, with folds seeming to swirl in every direction. This is particularly well seen at [NC 0581 2525] on a vertical face, when the tide is low (Fig. 3.21). Here, however, the

Figure 3.22 Achmelvich: basic dyke cutting gneiss, with gneiss xenolith inside dyke.

gneiss structure is now very steep, and this locality is just at the edge of the Canisp shear zone.

From here follow the footpath towards Alltanabradhan, noting on the way a superb example of a narrow vertical basic dyke cutting the gneiss bands [NC 0592 2509], just beneath a power pylon (Fig. 3.22). The dyke has a narrow sheared margin with a strong fabric of schist, but the interior has a random igneous texture, and contains xenoliths of gneiss, proving that the dyke is younger, as does the cross-cutting margin. On the opposite side of the track there is a shear zone in the gneiss parallel to the edge of the dyke. Many of the hornblende pods and balls in the gneiss have been weathered out to leave deep hollows. A very wide dyke crosses the path where it reaches its highest point [NC 0606 2551]. In most places it has a metamorphic fabric – note the white elongate streaky feldspar crystals in a very dark green matrix – but there are also patches where the original random igneous texture can be clearly seen (Fig. 3.23). From the top of this dyke there are good views to Lewis, and to Suilven, Cùl Mòr, Quinag, Stac Pollaidh, then the hills of Wester Ross. The nearby peninsulas (the farthest one is Rubha Leumair) clearly show the steep foliation of the gneisses within the Canisp shear zone. Additional exposures around here are described in Goodenough and Krabbendam (2011), while Tarney (in Barber *et al.*, 1978) describes the rocks on the south side of the bay; access

Figure 3.23 Achmelvich: basic dyke showing random igneous texture in centre of dyke (white feldspar and black hornblende).

above the caravan park is difficult on account of the dense undergrowth. Return to the car park, then drive to the B869, turn right and continue for 2km before joining the A837. On approaching Lochinver, note the vertical, rusty orange-weathering schists along the roadside, marking the edge of the Canisp shear zone.

Chapter 4

Lochinver to Assynt

Geological overview

The geology of Assynt has been studied intensely for over 150 years, and the district is a classic area for investigating complex systems of folds and thrusts at the edge of a mountain chain (Fig. 1.4). Rock exposure is excellent, and the visitor is rewarded with magnificent mountain and coastal scenery, where the link between landscapes and rocks is more obvious than in any other part of the British Isles. It was in Assynt that thrust faults were first identified and used in the interpretation of the sequence of structural events involved in mountain building. Here there is abundant evidence that large segments of the crust were transported northwestwards during the evolution of the Caledonian mountain chain. The geology of Assynt also played an important role in helping to explain the growth of other mountain chains, in particular the Alps. It is a superb outdoor laboratory, and has been used in the training of generations of geologists. One of the major attractions in this relatively small area is the great variety of easily recognizable rock types, spanning a huge portion of geological time, and the good exposure allows for accurate mapping in a reasonably accessible area. There are few other places in the world where it is possible to gain a three-dimensional impression of the edge of a mountain chain.

The oldest rocks are the Lewisian gneisses, around 2900 million years old, which were intruded by basic dykes at 2400 and 2000 million years ago, then refolded and partially melted, with the formation of granites and pegmatites at 1750 million years ago. Over the next 550 million years, these rocks from the lower crust were gradually uplifted as the overlying rocks were eroded away, to appear at the surface by 1200 million years ago. At that time in Earth history, Northwest Scotland formed part of Laurentia, a large continent that also included present-day North America and Greenland. The climate was arid, and land plants had not yet evolved. Erosion of rocks lying to the west and transport by large intermittent rivers brought large amounts of sediment in the form of sand and pebbles, laid down in alluvial fans, to produce the red and brown Torridonian rocks that form the spectacular mountains of Suilven, Stac Pollaidh, Ben More Coigach and Quinag. The junction between the flat Torridonian sedimentary rocks and the underlying folded and metamorphosed Lewisian is an unconformity, representing a time gap of 550 million years. It is superbly exposed in many parts of Assynt and represents one of the best examples of its kind anywhere. Earth movements caused the earlier rocks to be tilted by about 20°, and subsequent erosion and drowning by a shallow sea led to the formation of another, equally spectacular, unconformity. Pure white sand was deposited on top of the Torridonian and Lewisian rocks during the Cambrian period, about 500 million years ago. It was at this time the first animals with hard shells began to evolve, and some of their fossil remains can be seen in the Assynt district. These clean, shallow marine sands were compacted and cemented, and now form the quartzites that cap many of the hills in the northwest, with their shining white scree slopes – Foinaven, Arkle and Beinn Eighe, for example. The Cambrian quartzites were then overlain by carbonate rocks during the Ordovician period, about 450 million years ago, culminating in the Durness Limestone, a rock that plays an extremely important role in the way the landscape has developed. At the present day, the Torridonian is once again horizontal, and the overlying Cambrian to Ordovician rocks dip 20° to the east, indicating a reversal of the tilting that had affected the Torridonian rocks after they were deposited. Apart from this relatively gentle tilting, the sequence of rocks to the west of Kylesku and Inchnadamph has not been affected by later movements, and is a stable area known as the 'foreland' to the Caledonian mountain belt lying to the east. However, it is worth noting that some important Caledonian structures, such as the Outer Isles thrust, occur farther west.

Immediately east of the main road between Kylesku and Ullapool lies the junction between the foreland and the folded and faulted margin of the mountains. In this area lie the Moine rocks, a group of metamorphosed schists, originally mud, shale and sand, that are older than the Cambrian Quartzite and Durness Limestone, and about the same age as the Torridonian (i.e. 1000 million years), but placed on top of all these. Some 430 million years ago, the Moine rocks were physically transported westwards during the final stages of the Caledonian mountain-building period in a series of thrust faults. At the edge of the mountain chain, the sedimentary rocks lying to the west were compressed, folded, sliced up into thin slivers and piled up one on top of the other by the bulldozer effect of the mountain building. The ultimate cause of this event was the collision of the northern Laurentian continent with Scandinavia and with the European continent to the south, and the closure of the intervening Iapetus Ocean. In total there are four major thrust faults, from base (oldest) to top (youngest) the Sole Thrust, the Glencoul Thrust, the Ben More Thrust and the Moine Thrust, each one carrying folded piles of rock known as nappes above them and responsible for a great thickening of the crust by repetition of the layers within the nappes.

At Loch Assynt the road runs along the Sole Thrust, which can be seen opposite Ardvreck Castle, and the multiple repetitions of the sedimentary layers can easily be traced on the hillside above the road (see Fig. 2.10). More or less simultaneously with the thrusting, several large masses of igneous rock were intruded into the sedimentary rocks, contributing to the doming up of the crust in the Assynt area. Erosion of this great pile resulted in an eastward-directed indentation of the Moine Thrust zone in Assynt, which has created a fortuitous window into the crust beneath, thus providing a superb section through the edge of the mountain front. Accurate age dating of the igneous rocks allowed thrust movements to be dated for the first time, since the intrusions spanned the periods of thrusting. This is another unique feature of Assynt geology.

Locality 4.1 Lochinver – Lewisian gneiss

Pale grey and white Lewisian gneiss forms the shore beside Lochinver church, but the details are obscured by lichen. A better place to see the Lewisian is at Baddidarrach [NC 0888 2275] on the north side of Loch Inver – very fresh large blocks can be found at the sea wall and around the car park at the pottery (with a view of the western buttresses of Suilven (789m), built of horizontal Torridonian Applecross Formation). Drive round to the harbour and the Culag Hotel (Fig. 4.1). Vertical walls of the old quarry at the harbour and close to the sports centre at [NC 0914 2204] show grey granulite gneiss with wispy streaks and thin lenses of black amphibolite and large pods of bright green ultrabasic rock.

Leave Lochinver and drive towards Loch Assynt on the A837, noting the vertical rusty weathering schists (metamorphosed iron-rich shales)

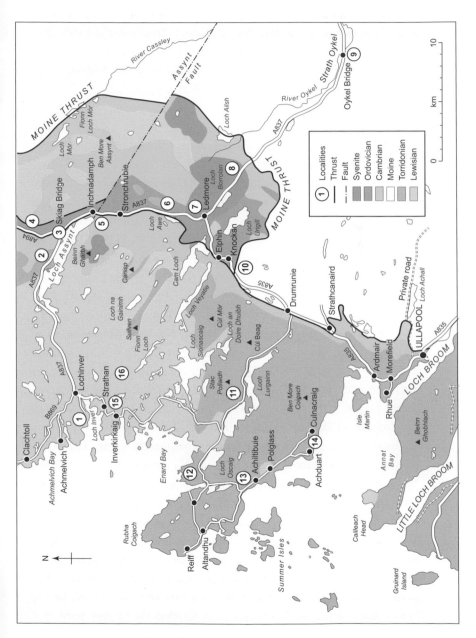

Figure 4.1 Route map with geology: Assynt.

on the left just outside the village, at the edge of the Canisp shear zone. Farther along the road is typical knock and lochan topography in the dark, coarse horizontally banded Lewisian gneiss, in many places covered in low mounds of hummocky moraine left by melting glaciers. At the lay-by [NC 1259 2439] there is a good view of Suilven to the south, and a superb view of the long ridge of Quinag, with the horizontal sandstone beds rising up above the Lewisian surface.

Locality 4.2 Loch Assynt

Park in the large lay-by (part of the old road) at the shore of Loch Assynt [NC 2121 2515] close to the clump of spindly Scots pines on a small island in a bay. Walk down to this bay to examine the Lewisian gneiss on the shore (it has green hornblende pods, streaks, elliptical lenses and balls, stretched out parallel to the foliation), where it is cut by a Scourie dyke that is soft and has been partly weathered out to form the bay [NC 2118 2510]. It has peridotite margins (ultrabasic) and a gabbro centre (basic). The exposures are smooth and rounded, and glacial striae (scratch marks) can often be seen. Above this, forming a prominent ridge above the shoreline, is a larger medium to coarse dyke with a random igneous texture internally, but with a definite foliation at the margins [NC 2103 2519]. The ridge can be traced for many metres, and the contact with the gneiss is sharp.

Now go back to the road and cross over to examine the low cuttings on the north side. Here [NC 2134 2515] the greenish Lewisian with steep banding is soft and deeply weathered and is overlain by flat Torridonian conglomerate (Diabaig Formation). This is the famous basal unconformity, representing a time gap of several hundred million years (Fig. 4.2). It can be examined at a number of places along this section [NC 2171 2510]. This may represent a fossil soil profile, i.e. the weathering may have taken place when the gneiss was at the surface, prior to the deposition of the pebble beds. However, the Torridonian is also weathered, and this may have occurred during the Cenozoic era, when deep chemical weathering took place everywhere. There is a good exposure of vertical banded gneiss beneath horizontal pebbly sandstone at [NC 2148 2514]. Some of the quartz pebbles are triangular in profile and may be wind-faceted, having been shaped on the desert floor by sand blasting, then transported by a river [NC 2128 2519]. If this section is walked out to the east, the level of the Lewisian–Torridonian boundary can be seen to rise and fall, indicating

Figure 4.2 Loch Assynt: Torridonian conglomerate and sandstone unconformably overlying mound of weathered Lewisian gneiss (bottom left).

pre-existing topography. At [NC 2194 2497] thick beds of sandstone in the Applecross Formation lie on top of thinner Diabaig gritty beds. Cross bedding, graded bedding and ripple marks are common in the Applecross rocks in this section, as far as the last exposures to the east.

Return to the car and drive east to the road junction at Skiag Bridge, noting on the way the smooth slope of the hill running up to the summit of Quinag. This bare rock pavement is a bedding plane of Cambrian Quartzite, unconformable on top of Torridonian rocks, which are horizontal here. Quartzite can be seen at the roadside just before the junction.

Locality 4.3 Skiag Bridge

Park near the A837/A894 road junction by Skiag Bridge, then walk left up the road for a few metres to examine the low cutting of pinkish quartzite [NC 2349 2440]. This is the Pipe Rock of Cambrian age, the pipes being the vertical traces of worms that lived in burrows in the beach sands (Fig. 4.3A, B). No actual body fossils are found, hence these are referred to as trace fossils, and have been called *Skolithos*. They are circular in cross-section and measure 1cm x 20cm approximately. Compare this with the situation beneath the Moine Thrust at the Stack of Glencoul (locality 3.16),

Figure 4.3A Cambrian Pipe Rock (quartzite), showing characteristic vertical worm tubes (trace fossil); Skiag Bridge, Loch Assynt.

Figure 4.3B Bedding plane view of circular tubes.

where the cross-section is more than five times that, and highly elliptical instead of circular. Good examples of the larger *Monocraterion* trumpet burrows – circular depressions with a raised point at the centre – can be seen on clean bedding planes in the Allt Sgiathaig burn, west of the road, at [NC 2304 2549].

Return to the road junction and walk in the direction of Ardvreck Castle. Low cuttings of distinctive brown and yellow to orange Fucoid Beds can be seen at [NC 2359 2423]. Looking in this direction from west of Skiag Bridge, it is obvious that these beds are lying on top of the Pipe Rock, so they are younger. In cross-section it is possible to see fine wavy laminations and cross bedding, while on the weathered top surfaces of the beds there are many examples of irregular interfingering markings that represent burrows and feeding trails. Originally these were thought to be seaweed, hence the name Fucoid Beds. Please do not hammer these surfaces. There are plenty of broken fragments in the cuttings. Note the blue colour of the fresh rock, in contrast to the brown weathered surfaces. Ripple marks can also be seen on the bedding planes. To the east, after a short gap, the pale grey Salterella Grit forms the next face [NC 2366 2414], opposite a small peninsula jutting into Loch Assynt, followed immediately by green grassy ledges of Durness Limestone, near the lay-by at [NC 2372 2406]. The Salterella Grit is a cross-bedded quartzite and has small (2–3mm) cream-coloured conical shells on the weathered surfaces. From here it is possible to see the folds in the hills above Inchnadamph (Fig. 4.4). Along this section, thrusting has produced 20 repetitions of the beds.

Figure 4.4 Loch Assynt: folds above Sole Thrust, which runs along loch shore at road level.

Locality 4.4a Assynt viewpoint

Stop at the viewpoint on the roadside going north from Skiag Bridge, at the P sign on the left of the road, uphill from Loch Assynt [NC 2343 2483]. Looking across Loch Assynt to Beinn Garbh, the horizontal beds of Torridonian can be clearly seen above the knobbly and irregular Lewisian landscape, then to the left (east) the dipping beds of white Cambrian Quartzite lie above the Torridonian and eventually cut across the Torridonian to rest directly on the Lewisian. This is the famous 'double unconformity'. The quartzite was deposited as sand in flat layers on a beach but is now tilted. The Torridonian was also originally laid down flat on an irregular surface of Lewisian hills and valleys, but was then tilted to the west and eroded before the quartzite was laid down on top. Later, the Torridonian was tilted back again to the horizontal, coincidentally, and now it is the Cambrian that is tilted 20° to the east (Fig. 4.5).

Locality 4.4b Quinag

Continue north past the grey mountain of Glas Bheinn (quartzite, above the Glencoul Thrust) with its impressive corrie and solifluction lobes on the steep slope – the thin soil has been slipping downhill under gravity on the smooth quartzite. At road level there are several exposures of brown

Figure 4.5 'Double unconformity' at Loch Assynt: horizontal Torridonian rocks in centre (high ground) resting on older knobbly Lewisian Gneiss (lower right); Torridonian overlain by younger Cambrian rocks, sloping down to the left.

Fucoid Beds and Durness Limestone, repeated by small thrusts. On the left, the glacial pavement climbing up to the top of Spidean Còinich on Quinag is a bedding plane of white quartzite. At the top of the hill there is a good view of the red horizontally bedded Torridonian sandstone forming the buttresses of Quinag, specifically Sàil Gharbh and Sàil Ghorm, in contrast to the white quartzite that forms the narrower and rounded summit of Spidean Còinich (Figs 2.10 and 4.7). Torridonian slabs occur on both sides of the road. After Loch a' Ghainmhich (Gaelic: sandy loch) the road twists and descends steeply. Park in the lay-by at [NC 2403 2956], facing Quinag. The pink rock here is a tough and brittle, sugary, even-grained quartzite (the Basal Quartzite unit of the Cambrian), showing clear cross bedding and frequent pink feldspar fragments among the white and colourless quartz. The feldspar was probably derived from local Torridonian and Lewisian sources. At the far end of the lay-by, under the waterfall that comes from Sandy Loch, there are several tall pinnacles of quartzite that look like isolated chimneys. Here the rock is badly shattered, being in a thrust zone; the pinnacles were formed when the loch drained rapidly into the deep overflow channel at the time of ice melting. It is possible to walk from here along the road to see the quartzite on top of the Torridonian at [NC 2402 2976], then east across the moor in the direction of Loch Glencoul to find the quartzite directly on Lewisian at [NC 2467 3017]. The Torridonian has been overstepped and is cut out by the Cambrian: this is the famous double unconformity again, mentioned at Skiag Bridge. The Torridonian–Lewisian unconformity can be seen about 500m west of here, towards the road, and is marked by the distinct break in slope. Beyond this point, the road northwards crosses knobbly Lewisian gneiss terrain, which can be seen in the distance north of Kylesku. Return to Skiag Bridge and turn left onto the A837 and carry on past Loch Assynt to Inchnadamph, 3km from the junction. The long ridge of Quinag is well seen on the Lochinver road (Fig. 4.6).

The road more or less follows the line of the Sole Thrust (most of the way cutting Fucoid Beds), and in some of the road cuttings it is possible to see how the dark limestones are steeply tilted and shattered. At Inchnadamph, note the outline of the loch, which has been gouged out in a northwest–southeast fault hollow. There is also a good view of the dip slope of quartzite coming down from the summit of Quinag. At Inchnadamph, there is a monument to Peach and Horne overlooking Loch Assynt [NC

Figure 4.6 Quinag, Assynt, near Lochinver. Horizontal Torridonian rocks above Lewisian gneiss basement.

2481 2222] and the Stronchrubie cliffs behind the hotel. Here the Durness Limestone has been sliced up by thrusting to form an imbricate structure (locality 4.5b). On the hill above the house opposite the telephone kiosk at Inchnadamph is a thick slice of Lewisian gneiss above the Glencoul Thrust.

Locality 4.5a Traligill caves

Park at the Mountain Rescue Post car park [NC 2510 2161] and walk up the footpath past the hotel and lodge, through the stile and to the valley of the Traligill river. The first exposures on the track opposite the river are pale Durness Limestone, belonging to the Eilean Dubh Formation. On top of the beds it is possible to see examples of clints and grykes, with plants growing in the protected hollows. A few metres farther on [NC 2576 2200], the limestone is now much darker and mottled, with many fractures. This is the Grudaidh Formation, which is older than the Eilean Dubh, hence there is a small thrust between the two. Note also the drystone walls around the fields, made of Durness Limestone and Fucoid Beds, with a mixture of the two rocks practically on the geological boundary. On the hill above on the left, small thrusts bring Lewisian gneiss over Cambrian Quartzite, which in turn is thrust over Durness Limestone. Along the path there is a good view of Conival on the skyline, consisting mainly of quartzite repeated several times over – individual beds are relatively thin, but the mountain side is 1000m high. A small patch of Torridonian is found towards the top, above the Ben More Thrust.

Walk past the house (Glenbain cottage) on the hill above the Traligill valley, and at the fork take the lower path on the right that goes towards the river (the main path climbs higher). Note the large glacial erratic block of Lewisian gneiss at [NC 2665 2127]. At Ruigh an t-Sagairt [NC 2672 2130] is the impressive outcrop of a thrust plane, the Traligill Thrust (Fig. 4.7). From here walk on to the white limestone pavement, an excellent example of karst weathering. The river disappears underground at a cave [NC 2706 2089], and follows the thrust. Note the vertical joints and strong orange-red coloration above the thrust on the opposite bank, where there is a dark (older) limestone that has been thrust over the white limestone. The section of the river below here is a dry valley, a typical feature of limestone country.

Cross a small wooden footbridge above the cave entrance and continue on the footpath up the valley. In spring and early summer, the area is covered in primroses, mountain avens and orchids. Another large cave, with a chimney (swallow hole) and disappearing stream are seen by the path at [NC 2758 2060], also within the Traligill Thrust. Take great care around here, and do not enter the caves. Notice folds in the limestone at a small dry cave at [NC 2761 20670]. This excursion can be extended

Figure 4.7 Traligill Thrust, Assynt; overthrust blocks of Durness Limestone; Quinag in distance, with Spidean Còinich on left.

95

along the valley towards the bealach and beyond – see Goodenough and Krabbendam (2011). Return by the footpath back to Inchnadamph.

Locality 4.5b Stronchrubie cliffs

Drive south on the A837 for 1.5km and park in the lay-by opposite the Stronchrubie cliffs, where there is a SNH information board concerning the glacial deposits (isolated moraine mounds) in the valley of the River Loanan that flows into Loch Assynt [NC 2482 2006]. The cliffs here, up to 100m high, show a clear example of small thrusts that push up the beds of limestone into steep angles (Fig. 4.8). Between the thrusts, the limestone beds are folded. One third of the way up from the base, there is a pale horizontal igneous sill, intruded into the darker limestone beds. At the top of the hill above the cliffs is a wide limestone pavement, 300m above sea level, part of the Inchnadamph National Nature Reserve. The plant communities here are exceptional in their variety. Look towards the U-shaped hollow of Loch Assynt from here, excavated by the ice, and to the quartzite-capped peak of Quinag.

Figure 4.8 Stronchrubie cliffs, Inchnadamph, showing small thrusts in limestone (steep features at top left).

Locality 4.5c Bone Caves

Continue south on the road for 2km to reach the car park on the left for the Bone Caves walk [NC 2528 1800]. This excursion follows the Allt nan Uamh (Gaelic: stream of the caves) to a line of caves on the hillside of Creag nan Uamh (Fig. 4.9). The small waterfall at [NC 2558 1794] occurs where an igneous sill intruded and baked the Salterella Grit. On the path to the left, exposures of pale yellow Durness Limestone can be seen. A bright green mossy area at the base of a small limestone cliff [NC 2608 1773] is caused by water gushing from the natural spring of Fuaran Allt nan Uamh (fuaran, Gaelic: a spring; Fig. 4.10A). The yellowish deposits here are due to calcium carbonate being formed at the surface as the water evaporates. This type of limestone is referred to as tufa. Upstream beyond here, the boulder-strewn river valley is dry in summer, except after prolonged heavy rain (Fig. 4.10B). Another spring is the Elephant Trap at [NC 2641 1741]. Continue on and take the obvious stone path on the right across the dry valley, leading steeply up to the caves at [NC 2681 1703]. Notice some folds in the limestone beds above and to the right of the cave

Figure 4.9 Inchnadamph: bone caves eroded into Durness Limestone.

Figure 4.10A Inchnadamph: spring emerging from karst limestone.

Figure 4.10B Inchnadamph: dry valley in limestone, path towards bone caves.

entrances. From left to right the caves are Fox's Den, Bone Cave, Reindeer Cave and Badger Cave. This is a scheduled Ancient Monument and Site of Special Scientific Interest, so there can be no hammering or collecting.

The caves, which are all shallow and filled with sediment, started to form around 200,000 years ago by dissolution of the limestone, when the level of the valley was much higher during the last glacial period. In 1889, Peach and Horne did some excavating and found animal bones buried in the sediment, including brown bear, arctic fox, wolf, reindeer, lynx, hare, vole, mouse, frog, lemming and many bird species. A polar bear skull dated at 19,000 years old is now in the Museum of Scotland in Edinburgh. The bones were probably washed into the caves during the melting phase of the last glaciers. Human bones from Reindeer Cave, dated at about 4500 years, suggest that the caves were used as burial places. From the entrance to the caves there is a view down the valley to a number of circular water-filled solution hollows about 1–3 metres in diameter, and across to the hillside opposite, showing a large landslide of Torridonian above the limestone. Since the Torridonian is older, it must have been thrust into its present position by the Ben More Thrust.

Return to the car park by the same path, then drive south for 2km and park in a lay-by on the shore of Loch Awe [NC 2495 1578], opposite a small quarry. To the west is the dip slope of Cambrian Quartzite on Canisp (846m), with Suilven visible to the left (Torridonian, 731m).

Locality 4.6 Loch Awe quarry

Cross the road and enter the quarry, which has large piles of fresh, broken material at the entrance. There is no need to approach the faces. Rusty-weathered brown–orange shales, the Fucoid Beds, have been quarried here, previously for possible use as a slow release potash fertilizer and subsequently for road material. Note the blue-grey colour when fresh – the rocks are hard and brittle, with finely laminated and more massive beds alternating. In places the bedding looks rather wavy or wispy (Fig. 4.11). Rare trilobite fossils (mostly the round head shield) have been found in this quarry, belonging to the species *Olenellus lapworthi*.

Locality 4.7 Ledmore marble quarry

Continue south for 2.5km, and park at the entrance of a large disused quarry, noting on the left the large block of green, yellow and cream Ledmore marble (Fig. 4.12). The quarry was in use for nearly 200 years. The marble is now used for road chippings, but previously it was cut and polished for ornamental use (mainly fireplaces – it was even exported

Figure 4.11 Loch Awe, Assynt: Fucoid Beds in roadstone quarry.

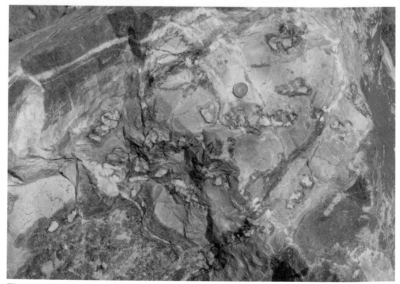

Figure 4.12 Ledmore marble quarry: thermally altered dolomitic limestone (dolostone), showing green bands of diopside.

to Italy, the home of marble). It formed as a result of metamorphism of Durness Limestone by heat and fluids from the nearby igneous intrusion (syenite) that forms the distinctive red hill above the road junction (Cnoc na Sròine). Small exposures at the roadside just south of the quarry show dark red syenite and marble side by side, but the location is too close to the road to be safe.

Locality 4.8 Aultivullin quarry

Drive past the Ledmore junction and past the Altnacealgach Inn on the left and Loch Borralan on the right. Note the many deep-red boulders of syenite strewn on the hill near the inn. Park just off the road at the forest clearing and walk across the rough ground into the disused Aultivullin quarry at [NC 2877 0968]. This has exposures of a type of syenite with the local name of borolanite. The quarry is important in the geology of Assynt for giving a reliable date to the thrust movements. The rock is grey overall, consisting of white potassium feldspar, grey nepheline and a distinctive dark mineral: melanite, a black to dark brown garnet. Quartz is absent, and this is referred to as an intermediate igneous rock. Xenoliths of dark fine-grained pyroxene-rich rock are common, and the greyish-white patches of nepheline are mostly flattened into elliptical shapes. A few narrow shear bands cut the rock. These facts indicate that the syenite was affected by thrust movements. However, there are also bands of coarse material that cut across the flattened areas, showing that igneous activity continued after thrusting. Age dates from here indicate that the last movements on the Moine Thrust took place 435 million years ago (Goodenough *et al.*, 2013).

Locality 4.9 Oykel Bridge

A detour can be made from here to Oykel Bridge [NC 3859 0087], 12km to the southeast, on the A837. This locality is famous for the superb mullion structures in Moine quartz-rich schist. They resemble a pile of giant flattened cigars, and are thought to have developed parallel to the main stretching direction during Caledonian deformation of the Moine rocks, in association with westerly directed movements on the Moine Thrust. The mullions can be viewed from the bridge, and there is no need to descend to the river bank. This is a Site of Special Scientific Interest, and no hammering is allowed.

Return to the Ledmore junction and turn left onto the A835; from the junction there is a good view of Cùl Mòr with its deep corries and twin peaks of Cambrian Quartzite outliers on top of Torridonian sandstone.

Locality 4.10 Knockan Cliff

Drive past Cam Loch through the village of Elphin, surrounded by green grassy fields on limestone, and with good views to Suilven and Canisp, then to the SNH visitor centre at Knockan on the left of the road facing Cùl Mòr [NC 1877 0908]. The white rock in the cutting by the entrance is Pipe Rock. There is a large car park and toilets at the centre, with information leaflets on this site and the Northwest Highlands Geopark. Knockan Cliff is an internationally important site – it is a protected SSSI and National Nature Reserve, and no hammering or collecting are allowed. At the start of the trail is a wall made of polished sections of all the local rocks in their correct geological order, and labelled. The flag-stones at the start of the trail come from Caithness (Devonian in age), but many of the steps on the path are from local white and pinkish Pipe Rock. Follow the clear footpath to the information hut, which has a specimen of a trilobite fossil (locality 6, Loch Awe). Beyond are labelled outcrops of Fucoid Beds and Salterella Grit. Towards the top of the trail is the famous outcrop of the Moine Thrust, with dark grey slabs of Moine mylonite thrust on top of the younger creamy Durness Limestone (Fig. 4.13). The limestone is contorted and fractured immediately below the thrust.

Figure 4.13 Moine Thrust at Knockan Cliff: grey Moine mylonite at top, above sheared yellow Durness Limestone.

Continue to the top of the hill above this point to Cnoc an t-Sasunnaich, for magnificent views to Cùl Mòr, Canisp, Cùl Beag, Conival, Ben More Assynt, Stac Pollaidh (locality 11), Suilven (locality 16), An Teallach and Ben More Coigach. The green sward of Elphin is impressive, with the limestone cliffs above the Sole Thrust, and the Ledmore marble quarry (locality 7) close to the red hill, Cnoc na Sròine, that marks the Loch Borrolan intrusion (locality 8). The Ben More Thrust goes up over the Breabag Dome above Elphin, marking the so-called Assynt Culmination within the thrust zone. These features are well displayed in photographs and cross-sections on the 1:50 000 Assynt geological map (BGS, 2007).

Take the path along the top of the hill in a southerly direction, crossing Moine mylonite. On the left, just before the path starts to descend [NC 1887 0878], there is a vertical wall of flat, light grey mylonite that can be examined (Fig. 4.14). There are tiny folds in the mylonite banding here, parallel to the cleavage. At first glance, the structure looks like bedding – which explains Murchison's mistake in calling these rocks Silurian, as they lie directly above the Ordovician Durness Limestone on the path just below. The mylonite is very thick above the base of the thrust, a clear indication of a large amount of strain during thrust movements. Return to the

Figure 4.14 Section through horizontal Moine schist mylonite above Moine Thrust, at top of Knockan Crag trail.

car park by the footpath. From the car park, note the view across to Cùl Beag and Cùl Mòr, the gap between them being a trough cut by moving ice that was also responsible for giving Stac Pollaidh, seen in the distance, its streamlined shape. The valley floor is largely blanket bog on top of poorly drained quartzite, which produces an acid soil.

If there is time, it is worth parking in the lay-by across the road from the Knockan entrance, then go through the gate on to the walkers' footpath that leads to Cùl Mòr, for a view of the thrust plane on the cliffs opposite. There are excellent examples of Pipe Rock on the path at [NC 1886 0961].

Travel south (the road follows the Moine Thrust – note the dark flaggy schist in the cuttings) for 4km to the junction with the unclassified road that leads to Achiltibuie. This is a narrow single-track road that twists its way along Loch Lurgainn, past Cùl Beag on the right and Ben More Coigach on the left, with its impressive five peaks of Torridonian sandstone. The castellated ridge of Stac Pollaidh (613m) soon comes into view (Fig. 4.15A).

Locality 4.11 Stac Pollaidh

Park in the walkers' car park at the foot of Stac Pollaidh (Stack Polly) [NC 1077 0949]. Cross the road and go through the gate then follow the path, crossing nearly horizontal pebble beds in Torridonian sandstone.

Figure 4.15A Stac Pollaidh: horizontal Torridonian beds.

The round pebbles are of quartzite, pegmatite and granite from Lewisian sources. Cross bedding is a distinctive feature in many places (Fig. 4.15B). Continue on the path round the shoulder for views to Suilven, Cùl Mor, Cùl Beag and Ben More Coigach. Red scree on Canisp comes from shattering of an igneous sill, the Canisp porphyry. Note how Suilven, like Stac Pollaidh itself, has a streamlined shape, caused by ice moulding. The landscape of isolated Torridonian sandstone mountains on top of the bare knobbly Lewisian pavement is most impressive. To the west can be seen the irregular, indented coast, and the northwest-to-southeast grain of the land, reflecting the structure of the gneiss basement – folds, shear zones, faults and dykes. Knock and lochan topography is very obvious from here; note too how the northeast–southwest fractures have resulted in the land being broken up into a series of squares. Loch Sionascaig occupies one of several large glacial troughs, all of them oriented southeast to northwest, reflecting the movement direction of the ice. Climb to the top for a view of the castellated ridge that results from frost shattering in vertical joints and along bedding planes. Scree material tumbling from the top has created a beautiful apron on the lower slopes, mostly stabilized by heather and bracken, but fresh rivulets of scree (also known as talus) demonstrate that the process is still active.

Figure 4.15B Cross bedding in Torridonian sandstone; footpath to top of Stac Pollaidh.

Return to the carpark and drive for 9km to the northwest past Loch Bad a' Ghaill and Loch Osgaig on the left, and park at the south end of Achnahaird Bay, in the large lay-by on the right where the road forks (Achiltibuie left, Reiff right), at [NC 0211 1242].

Locality 4.12 Enard Bay

Walk up the road from the car park and go left at the bend in the road [NC 0245 1270], heading for the rocky knoll on the skyline (to the north), provided only if the weather has been settled and the peat moor is dry. Otherwise it is best to walk round the east side of Achnahaird Bay, with red sandstone outcrops of the oldest group of the Torridonian (Stoer Group), displaying cross bedding on a large scale (e.g. at [NC 0223 1418]), which could indicate either a fossil sand dune or a river channel. Walk over the hill of Cnoc Mòr an Rubha Bhig, which is an ice-rounded outcrop of coarse Lewisian gneiss and pegmatite, with patches of Torridonian conglomerate on top. Walk down to the shore in the bay of Camas a' Bhothain (Gaelic: bothy bay), paying due attention to the numerous deep potholes on the hillside, resulting from an attempt to reforest the slopes. The holes are filled with water and overgrown with heather. Low tide is best for the coastal out-crops at this locality. A ruined salmon bothy at [NC 0285 1465] provides a convenient reference point. On the shore just west of the bothy at [NC 0277 1465] there is a dark red mudstone, cross-bedded and with ripple laminations, containing some large boulders of Lewisian gneiss. Around the boulders the bedding has been disrupted, probably due to slumping down the dip slope of an alluvial fan in a river system. Another interpreta-tion, that these represent blocks that fell from the base of an iceberg, has not gained favour. Walk round to the other side of the small bay to the east of the bothy, noting the sedimentary breccia of gneiss, to a very obvious dome at a pronounced inlet [NC 0284 1464]. On this dome, which is made of Lewisian gneiss, there are boulders of gneiss and red sandstone, surrounded by finely laminated red limestone (Fig. 4.16). Algal growths were responsible for the structure, which is similar to stromatolites, so these represent primitive life forms that built up sheets, films and mats in a shallow lake directly on top of the gneiss. Please do not damage this unique exposure. Bright red mudstone occurs above the limestone, followed by a conglomerate containing very large blocks that may represent parts of a cliff face that slumped and fell.

Figure 4.16 Enard Bay: Torridonian algal mudstone draped over Lewisian gneiss knoll, with angular gneiss breccia cemented by the limestone. Note Suilven in background.

Continue round the bay to the east and go down onto the wave-cut platform either by a series of small ledges, or round the top and down to the left. Look back to the small cliff for a view of the boulder conglomerate. On the bare slabs here [NC 0302 1468], the red sandstone has numerous pea-size lapilli and dull green feldspar crystals, altered to clay – this is the Stac Fada volcanic member in the Stoer Group (Table 2.2), seen at Clachtoll (locality 3.18). In places the beds are disfigured by sample drillholes. Lapilli form when rain falling through a cloud of ash causes the dust to stick together. Many of them have a concentric pattern in cross-section (Fig. 4.17). This formation is also controversial, with the claim that it represents material ejected during a meteorite impact. See Amor *et al.* (2019), and Sims & Ernstson (2019) for a discussion of the opposing viewpoints. Now walk round to the east side of the headland of Rubh' a' Choin [NC 0340 1470]. Here the grey shales on the slabs belong to the Diabaig Formation at the base of the Torridon Group (1000 million years old), while the coarse red sandstones in the small cliff above are in the Applecross Formation.

From Enard Bay there is a view to the mountains in the east: Suilven, Cùl Mòr, Cùl Beag and Stac Pollaidh. Either return by the same route, being sure to avoid the treacherous bright green quaking bog in Bad Fhluich ('the wet hollow'), or go along the eastern side of the peninsula, walking across slabs of Torridonian (Applecross Formation), as far as Loch Garvie. Then take the track back to the road at the bridge by Loch Osgaig [NC 0390 1300] and return to the car park, about 2km away.

Figure 4.17 Enard Bay: volcanic ash lapilli in Stac Fada member.

Locality 4.13 Achiltibuie

Drive to the village (there are toilets on the left just before the village is reached) and park at the Summer Isles Hotel [NC 028 078] then walk south towards Polglass. Just before the school, a track between the buildings goes up the hill on the left to an obvious low craggy exposure, Creag a' Phuind, at [NC 0285 0838], about 350m northeast of the hotel. This is an inlier of Lewisian gneiss, surrounded by Torridonian, as can be easily seen from the landscape. From the hillside there are views to the Summer Isles, also Torridonian. The Lewisian rocks here are part of a layered basic to ultrabasic igneous intrusion, and these particular exposures have large clusters of deep red garnets in the basic units, together with coarse dark hornblende, especially well seen to the southeast in the small cliff at [NC 0325 0803] (Fig. 4.18). The ultrabasic units are peridotite, consisting of olivine and pyroxene arranged in separate mineral layers, possibly as a result of gravity settling as the intrusion cooled and crystallized. The rocks have an igneous texture, with the large even-shaped crystals arranged in random fashion, not streaked out like a gneiss. Feldspar-rich units form lighter-coloured layers near the top of the section. This particular body is similar in many respects to the one at Drumbeg (locality 3.17).

Return to the road and drive south for 4km then turn right towards Achduart and continue for another 1.5km until a cairn at the cliff top comes

Figure 4.18 Garnet–pyroxene gneiss, Lewisian, Achiltibuie; clusters of red garnet crystals surrounded by white feldspar reaction rims.

into view, then park off the road where it bends to the left [NC 0467 0417], near a house. The impressive backdrop of nearby Ben More Coigach, with its corries, is made of Applecross Formation sandstone.

Locality 4.14 Achduart

Walk across the heather on a sheep track to the mossy green knoll with a cairn [NC 0428 0380]. On the way, the flat slabs are pebbly horizons in coarse, cross-bedded red Applecross Formation sandstone, laid down by large rivers. Go towards the point of Rubha Dubh Ard and down to the broad flat bedding planes above high water mark. This is quite easy, but care must be taken, and this should not be attempted in misty weather. Just offshore is Horse Island, with Tanera More and Tanera Beg a little bit farther out in the bay. The grey and green slabs are made of finely laminated Diabaig shales, while by contrast the small cliff is of red Applecross sandstone, blocks of which have tumbled on top of the shales. At the junction, the shales are stained red by the overlying sandstone. The base is irregular, as a result of erosion, and the sandstone has flakes of the underlying muds at its base. Sedimentary structures in the shales include ripple marks and small crumples due to slumping (Fig. 4.19).

Return to the car and go back through Achiltibuie and past Stac Pollaidh to join the main A835 road at Drumrunie, then turn right and

Figure 4.19 Achduart: red Applecross sandstone above green and grey Diabaig shales and siltstones, a junction within the Torridonian.

go to Ullapool, 15km to the south. An alternative route from Lochinver is via the minor road south, to Inverkirkaig then Achiltibuie. If this route is chosen, then there are two additional stops.

Locality 4.15 Strathan

Park in the lay-by at [NC 0821 2058]. The prominent rocky ridge on the roadside by a bridge crossing the stream is a Scourie dyke. It is a fresh rock with a coarse, even-grained, random igneous texture. Minerals present are large dark red feldspars and black olivine and hornblende. Veins and patches of bright green epidote cut through the dyke in places, as a result of later alteration.

Locality 4.16 Kirkaig Falls and views to Suilven

On the Lochinver road there are good views to Stac Pollaidh, Cùl Mòr and Suilven. Drive south to Inverkirkaig, twisting through knobbly Lewisian country. Stop in the car park where the road takes a sharp left turn, at the River Kirkaig, before the village [NC 0855 1934]. Follow the footpath signposted for the Falls of Kirkaig, 3km away, where the path forks to the right, for a view of the waterfall [NC 1114 1785]. To see the Torridonian–Lewisian unconformity at the foot of Suilven at its best, return to the path and continue for about 400m to [NC 1156 1775] where the path turns left. Suilven stands proudly above Fionn Loch, and from here there are views to Cùl Beag, Cùl Mòr and Stac Pollaidh as well. Note the deeply scoured, rocky Lewisian landscape and the streamlined shape of Suilven itself (Fig. 4.20).

Figure 4.20 Suilven: Torridonian sandstone resting on a basement of Lewisian gneiss (forming typical knock-and-lochan feature).

Chapter 5

Ullapool to Gairloch and Loch Maree

Introduction

This chapter focuses mainly on the Lewisian gneiss, particularly the folded, sheared and metamorphosed sedimentary and volcanic schists of the Loch Maree Group and the Scourie dykes, and the youngest sedimentary formations of Northwest Scotland – the Triassic (New Red Sandstone) rocks that occur intermittently along the west coast. The route passes An Teallach, one of the most majestic mountains built of Torridonian sandstone and famed for its outstanding variety of glacial features. The area lies within the Wester Ross National Scenic Area.

On the way to Ullapool from Assynt (Fig. 5.1), the A835 road more or less follows the Moine Thrust, as can be seen from the frequent road cuttings in dark green, black and grey almost flat slabs of mylonite and schist. Beyond Strathkanaird, the road follows a valley eroded along a fault, with shattered pink Cambrian Quartzite on the northwest side and brown Torridonian sandstone on the southeast. At Ardmair, Ben More Coigach comes into view, opposite Isle Martin, with the Summer Isles in the bay beyond (Fig. 5.2). Note the crescent-shaped spit on the shore at Ardmair, made of round white quartzite cobbles. The horizontal bedding of Torridonian (Applecross Formation) is magnificently exposed in the sheer cliffs of Ben More Coigach. These beds are well seen by the road-side just south of Ardmair, and if time allows, it is worth stopping briefly at the entrance to the track leading to the holiday cottages at Morefield. Examine any of the broad flat slabs of chocolate-brown coarse sandstone. Cross bedding, irregular slumped bedding and thin pebble horizons that grade upwards into sandstone abound, indicating that these rocks were laid down in a river environment. The slumping may have resulted from earthquake activity, causing wet sediment to slide down underwater slopes.

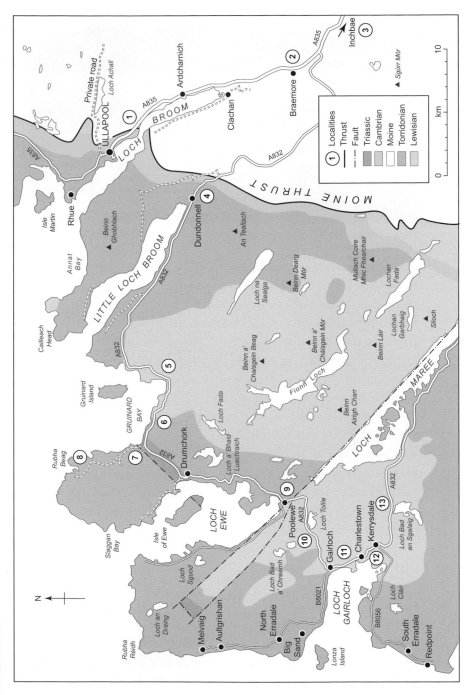

Figure 5.1 Route map with geology: Ullapool to Gairloch.

Figure 5.2 Ben More Coigach: view from Ardmair of horizontally bedded Torridonian rocks.

Other structures suggest that water was expelled and forced upwards under the sudden weight of additional layers of sand. Continue south towards Ullapool and stop for a panoramic view of the town as Loch Broom and the Fannich Mountains (of Moine rocks with Lewisian inliers) in the distance come into view. The town was built in a planned rectangular grid fashion (1788), on a post-glacial gravel terrace.

To the east of Ullapool, the Moine Thrust has an eastward indentation similar to the much larger one at Assynt (Fig. 4.1). Here it is called the Achall culmination, and extends to Loch Achall in the hills above the town. Within the culmination are Lewisian gneiss, Torridonian sandstone, Cambrian Quartzite and Pipe Rock, and Durness Limestone (which is being actively quarried for roadstone), folded and thrust together, then completely overthrust by Moine rocks above the Moine Thrust. The Moine rocks have been converted to mylonite, and the quartzites show intense brittle fracturing everywhere in the vicinity of Ullapool. Good exposures of white and pale pink quartzite occur in the roadside cuttings just south of the town, but these are not safely accessible, as the road is too busy.

Locality 5.1 Ullapool

Park as convenient in the town (always very busy in summer), take the marked footpath from the northern outskirts [NH 1300 9450], and walk about 1km uphill to the flat top of Meall Mòr [NH 1430 9463]. An alternative would be to park near the entrance to the Morefield private road [NH 1290 9498] that leads past the working quarry (in Durness Limestone, for roadstone), towards Loch Achall. Lorries use this road and care must be taken when walking there, especially near the weighbridge and loading bays (wear a fluorescent jacket if possible). Do not enter the quarry, although it is possible to look in at the entrance for a view of the dark grey and blue limestone dipping 45° to the east. If the first option is chosen, the rocks seen at the top of the hill are thinly laminated greenish-grey Moine mylonites above the Moine Thrust. The hills on the other side of the valley to the north are smooth bare crags of white Cambrian Quartzite, streamlined by the ice. Bedding planes dip at 10–15° upstream, forming a series of steps or small escarpments – this feature can also be seen clearly from the ferry terminal.

Either return across the bridge then down the road to Ullapool, or continue (down the geological succession) on the well-exposed quartzite and Pipe Rock beds to the west, crossing onto brown Torridonian sandstone near Creagan Dubha (Gaelic: black crags) at [NH 1351 9576], then down hill via a gentle slope to avoid the river gorge, and reach the A835 road at [NH 1229 9529], about 1km north of Ullapool. For a fuller excursion to the Achall Culmination, see Goodenough and Krabbendam (2011); the BGS map (Sheet 101E, 2008) has a detailed large-scale inset of the Achall area, and cross-sections through the thrust pile.

Locality 5.2 Corrieshalloch Gorge

South of Ullapool, Torridonian and Cambrian rocks are thrust over Durness Limestone north of Corry Point on Loch Broom, and to the south of this peninsula there are exposures of mylonites in the Moine rocks (Morar Group), with small-scale isoclinal folds affecting the thin lamination in these fine-grained rocks. The mylonites lie within the Moine Thrust zone. Some 16km south of Ullapool there is a large car park on the left, with toilets (previously this was the car park for the Corrieshalloch Gorge viewpoint, National Trust for Scotland). Flat slabs of pale grey quartz–biotite schist (pelite) interleaved with quartz-rich psammite can be examined here. There is a prominent stretching lineation on the foliation surfaces of the schists. Much of the strain associated with west–northwest movements on the Moine Thrust was taken up by the weaker pelites. Now continue for 1km to the Braemore Junction and turn right onto the A832 and park on the right after 750m in the NTS car park. Walk down the track to the bridge for a view of the Falls of Measach and the Corrieshalloch Gorge (a National Nature Reserve, and one of Scotland's 51 best geological sites), formed by overdeepening of a joint in the Moine schists by a meltwater stream flowing beneath a glacier. The rocks can be examined in the car park and on the low road cutting opposite. These are coarse, quartz-rich mica schists, cut by quartzite veins and pegmatites. See Kelley (in Strachan *et al.*, 2010) for more on the geology of this area.

At this point, a detour can be made to Inchbae to see the famous augen gneiss (or decide to carry on to Garve then Inverness). Return to the Braemore Junction, turn right and follow the A835 southeast for 20km towards Inchbae Lodge. Alternatively, this locality can be visited on the way north to Ullapool or Assynt.

Locality 5.3 Inchbae (optional detour)

Just to the west of Inchbae Lodge, stop at the road bridge (Black Bridge – An Drochaid Dhubh) over the Blackwater river by the small spillway of an old dam [NH 3735 7079]. Water-washed pink and white slabs can be examined under the bridge, but the blasted blocks on the west side show the rock texture rather better (Fig. 5.3). This attractive rock is the Inchbae augen gneiss, part of the Carn Chuinneag granite intruded 600 million

Figure 5.3 Inchbae augen gneiss at Black Bridge.

years ago into Moine schists, then deformed 470 million years ago in the Caledonian orogeny. (Another example of igneous activity of this age can be seen at Tayvallich (Chapter 9, locality 4), where there are some spectacular pillow lavas dated at 600 million years old.) In this augen gneiss, consisting of large crystals of quartz, feldspar, biotite and muscovite, the original large feldspar phenocrysts were flattened into eye shapes during deformation. Trails of this rock occur as ice-transported boulders eastwards from here to the Moray Firth, and since the rock is so distinctive, the boulders provide an important glacial marker for the movement direction of major ice sheets. Note the boulder-strewn landscape around here, with both Inchbae gneiss and pale grey Moine psammites (which show cross bedding). Some of the nearby hills have small Old Red Sandstone outliers on top of Moine schist.

If a decision is made to head south, then at Garve it is worth stopping in the car park (toilets) by the old bridge over the Blackwater [NH 4025 6391]. There are excellent, easily accessible, exposures in the river bed of flat slabs of Moine psammites, with the near-horizontal foliation affected by gentle open folds (Fig. 5.4). Closer towards Garve, the roadcuts show steeper foliation planes (50° dip) in silvery quartz-rich schists, adjacent to the Sgùrr Beag Thrust.

Figure 5.4 Garve waterfall: open folds in gently dipping Moine rocks.

To continue to locality 4, return to the Braemore Junction and drive northwest on the A832 road towards Dundonnell. On the way, note the view north from Loch Glascarnoch [NH 310 730] to Beinn Dearg, made of Moine psammite. The large stream (Allt Mhucarnaich) flowing down the mountain occupies a glacial U-shaped valley. After the junction, there is a large car park at [NH 1958 7843]. Stop here for a brief view down Strathmore towards Loch Broom, a flat-bottomed, U-shaped glacial valley filled with post-glacial sediments. Moine schists are seen on both sides of the valley. In places where sedimentary structures can be seen, such as cross bedding, the rocks can be shown to be right way up. From the start of the way-marked path to Kinlochewe near Loch a' Bhraoin [NH 1620 7600] there is a view into the Fannich Mountains, made of Moine schists interfolded with Lewisian gneiss inliers. On the hill to the north of the loch, Meall an t-Sìthe, there is an important structure, the Sgùrr Beag Thrust. See the Moine Guide for a challenging excursion here (Kelley, in Strachan *et al.*, 2010). This thrust separates Glenfinnan mica schists (pelites) from the older quartz-rich Morar psammites (see Table 2.4).

The steep cliffs along the Dundonnell river valley consist of dark, flaggy horizontal Moine rocks. An Teallach (1062m) comes into view from the ruin of Fain Bothy at [NH 1383 7931]. Note the extensive heather-covered moraine hummocks in this area. An Teallach is made of flat Torridonian sandstone (Applecross Formation) capped by Cambrian Quartzite, which makes up some of the rounded summits and is very obvious from the white screes (Fig. 5.5). The quartzite, which dips about 20° to the southeast, lies

Figure 5.5 An Teallach, Torridonian sandstone, with Cambrian quartzite forming white screes on right.

unconformably on top of the Torridonian and comes down to the road at Strath Beag near Dundonnel, where it is cut out by the Moine Thrust. An Teallach is an important site for studies of glacial phenomena, including corries with end moraines, glacial striae and erratic boulder trains. During the Loch Lomond Readvance 10,000 years ago, each of the six major corries on the mountain had its own small glacier.

Locality 5.4 Dundonnell

For an outstanding view of An Teallach, take the small side road on the right at Dundonnel House towards Badrallach, crossing over the Moine Thrust onto white Cambrian Quartzite (beds seen in the river, dipping to the east), then prominent slabs of horizontal red and brown Torridonian sandstone. The hillsides are stepped terraces of sandstone, littered with many large boulders. The hills to the west of An Teallach are Sàil Mhòr and Sàil Bheag, also Torridonian. Sedimentary structures can be examined in the slabs by the road, including cross bedding, graded bedding, pebble beds and slump structures (Fig. 5.6). Turn at the sharp bend in the road by the gate that marks the entrance to a track [NH 1005 9195].

Return to the main road and turn right, passing the Dundonnel Hotel on the shores of Little Loch Broom and in the shadow of the towering

Figure 5.6 Dundonell: slump folds and dewatering structures in Torridonian sandstone.

walls of An Teallach. The loch is a glacial fjord. Horizontal bedding of the Torridonian on the north side is very well expressed. A good view across to the twin summits of Beinn Gobhlach, dominating the Scoraig peninsula, is obtained from the large car park at the top of the hill before the bend [NH 0025 9280]. As soon as Gruinard Bay is reached at Mungasdale, there is a change in the landscape to the knobbly hills typical of the Lewisian, in sharp contrast to the smooth outlines on Torridonian rocks. Gruinard Island, which is made of Torridonian, illustrates this contrast very well.

Locality 5.5 Gruinard Bay

The unconformity between Lewisian and Torridonian rocks can be examined at Gruinard Bay. Park in the beach car park at [NG 9521 8988], then walk to Inverianvie Bridge at the bend in the road, noting on the way the shearing and steep joints in the cliffs by some trees at the bridge. Note also the Lewisian cliffs on the roadside as it climbs the hill, with the contrasting gentle landscape of Torridonian on the low peninsula to the north. At the bridge [NG 9510 8983] and in the cliff on the other side of the road, the base of the Torridonian (Stoer Group; see Table 2.2) is marked by a conglomerate of large round boulders of Lewisian gneiss that must have been transported by a very fast-flowing river (Fig. 5.7).

Figure 5.7 Boulder conglomerate of Lewisian gneiss clasts at base of Torridonian, Gruinard Bay.

Figure 5.8 Raised beach at Little Gruinard, with An Teallach in background (Torridonian above Lewisian basement).

Figure 5.9 Lewisian gneiss (agmatite) at Little Gruinard.

Continue past Little Gruinard up the steep hill to the parking place at the corner [NG 9403 9012]. Great care is needed here when entering and leaving this car park and when walking to examine the roadside exposures, because of the blind summit ahead. Across Gruinard Bay is a post-glacial raised beach, with An Teallach in the background, rising above the typical Lewisian knobbly landscape (Fig. 5.8). The road cuttings on both sides show good examples of black and white coarse-grained banded gneiss, with a mixture of nebulous (wispy) migmatite, amphibolite and agmatite (angular black amphibolite blocks surrounded and invaded by white quartz–feldspar veins and patches, Figure 5.9). This hill of gneiss has been exhumed by erosion from beneath the Torridonian, and was a hill when the Torridonian was being laid down, about 1000 million years ago. West of this viewpoint, the road descends gently, eventually returning onto Torridonian rocks.

Locality 5.6 First Coast

Park on the roadside between Second Coast and First Coast above a narrow inlet at [NG 9256 9104]. Grassy slopes lead to the sea, where there are large west-dipping slabs of Stoer Group rocks (low tide is needed). Red beds are exposed below the low cliffs at [NG 9241 9111]. These are coarse and gritty, with round pebbles of Lewisian gneiss that were transported by rivers and deposited on an alluvial fan. Exposures just below the parking place are in the Stac Fada Member, containing irregular green fragments of altered volcanic material, 1–2cm in size. The Stac Fada Member (also seen at Stoer and Enard Bay) represents ash material that fell onto the contemporary surface and was washed downslope by rivers. Being so distinctive, it makes a very useful marker horizon. From this point there are excellent views to An Teallach and Ben More Coigach.

Continue west on the A832 for 2km and turn right at Laide, then park near the old jetty after another 1km [NG 9017 9256].

Locality 5.7 Laide

Low raised beach sea cliffs and exposures along the shore at Laide show bright red and pink Triassic (New Red Sandstone) conglomerate with horizontal bedding. Boulders and seaweed rather obscure the rocks in places, and low tide is needed for the exposures on the foreshore. The conglomerate consists of round flat pebbles of Lewisian gneiss, Torridonian sandstone, Cambrian Quartzite and Durness Limestone cemented by white quartzite (Fig. 5.10). Many of the Torridonian pebbles are cut by thin quartz veins. At the boathouse the conglomerate grades up into finer light grey sandstone, with only occasional pebbly horizons, containing smaller clasts. Undercutting by the sea has produced caves in the pink sandstone and boulder conglomerate. The low cliff on the south side of the beach, towards the ruined chapel and graveyard, is made of bright red even-grained sandstone. These river-lain beds are the same as the Stornoway Beds on Lewis, and the Minch Basin–Sea of the Hebrides region is underlain by this formation, banked up against the Minch Fault that runs down

Figure 5.10 Triassic conglomerate, Laide.

the east coast of the Outer Hebrides. Fault movements during sedimentation allowed substantial thicknesses (up to 1000m) of sandstone and conglomerate to accumulate in the basin.

Continue northwards on this road for 4km to the small village of Mellon Udrigle and park in the large car park near the beach and the old schoolhouse [NG 8905 9600].

Locality 5.8 Mellon Udrigle and Rubha Mòr

The Rubha Mòr peninsula north of Mellon Udrigle is the type area for the Aultbea Formation of the Torridon Group (see Table 2.2). It lies above the Applecross Formation and is distinguished by its much finer grain size, i.e. sandstones and no conglomerates, and the very abundant contorted bedding structures resulting from the expulsion of water from underlying layers. Walk 1.5km north on the footpath from the car park, passing over some rather uninformative ice-smoothed, lichen-covered inland outcrops (very slippery when damp). On the coast at Rubha Beag the beds dip at 30° to the southeast. They are coarse dark red sandstones, showing many examples of cross bedding, ripple marks and contorted bedding due to slumping during deposition. The junction between the Aultbea and Applecross formations lies at Mellon Charles on the west side of this peninsula.

Locality 5.9 Loch Ewe

Return the same way and rejoin the A832, turning right at Aultbea. After 4km stop in the large lay-by and cross the road (with care) to the viewpoint on the corner above Rubha Thùrnaig, overlooking Isle of Ewe [NG 8716 8491]. The rocks here are Applecross Formation sandstones, with the hill above being made of Lewisian gneiss. Excellent views can be had south from here to the Torridonian hills of Wester Ross – Beinn Airidh Charr, then Ben Eighe and Coire Mhic Fhearchair. Nearby, note the smooth landscape of the Torridonian: gently curving bays, low hills, farms, fertile fields. Isle of Ewe is Torridonian, with a small area of Triassic in the south. On the west shores of Loch Ewe, the rocks at sea level are Lewisian, as can be clearly seen from the topography.

Continue towards Poolewe (take great care when leaving the car park), travelling down the Torridonian escarpment. Note the gently dipping slabs, covered in the characteristic white and light grey lichen on weathered surfaces, and the hillside scattered with countless sandstone slabs. Parking, toilets and restaurant facilities are available at the Inverewe Gardens (National Trust for Scotland). The gardens are on Torridonian rocks (Stoer Group), and from the paths there are good views across Loch Ewe to the typical rocky Lewisian landscape with much bare gneiss exposed. The Lewisian appears here as a result of faulting against the northwest-trending Loch Maree Fault, which is responsible for the straight edge of Loch Maree and the southwest shore of Loch Ewe (Fig. 4.1). The fault is a dextral strike-slip fault (vertical, with a right-handed movement), with a displacement of 15km. It separates the Loch Maree and Gairloch belts of Lewisian metasedimentary rocks. Further dextral movements occurred after the Moine Thrust was formed, since the thrust is offset by 2km by the fault.

After crossing the broad flat terrace south of Poolewe, the road climbs onto vertical Lewisian gneiss. Park on the corner in the large lay-by at [NG 8585 7895] near the minor road to Tollie Farm, 2km south of Poolewe. The hill on the right is Creag Mhòr Thollaidh. From here there is an excellent view along the Loch Maree Fault, with Beinn Airigh Charr (791m) on the left, made of Lewisian amphibolite.

Locality 5.10a Creag Mhòr Thollaidh

Take the clear track that leads to Slattadale, crossing a ford, through a gate and another ford. Note the smooth glaciated surfaces, streamlined ridges and perched boulders left by glaciers. The Lewisian rocks here are various types of gneiss, pegmatite and granite. The footpath follows the line of a fault, and in many places the gneisses are severely crushed and injected with black pseudotachylite veins, especially well seen on a bluff to the left of the path beside a small wooden bridge [NG 8600 7755]. Pseudotachylite forms when gneiss is rapidly melted during brittle fracturing caused by fault movements. It is essentially a glass, injected as a liquid into numerous cracks within the fault zone. The bright green veins are epidote, an alteration product of hornblende. An age date of 1000 million years for this rock indicates that it may be related to the Grenville orogeny. Shortly after this point, the path becomes narrower and steeper, especially at the small waterworks, crossing bare exposures of red pegmatite and quartz–feldspar gneiss. Carry on up the path for another kilometre and cross to the hill on the left side of the valley after passing a loch on the right [NG 8654 7675]. This hill is made of a large Scourie dolerite dyke with a random igneous texture internally (i.e. undeformed) and a planar fabric (shearing) at the margins. It is folded and cuts black, grey and white coarse banded gneiss that is much more deformed than the dyke, indicating that the dyke and host gneisses were affected by later events, related to Laxfordian deformation. The gneisses are strongly folded and contain elliptical amphibolite pods and lenses parallel to the banding and foliation, and are criss-crossed by granitic veins that formed by partial melting of the gneisses during the Badacallian event.

Return to the car and drive west for 1km, opposite the east end of Loch Tollaidh (take care when leaving the lay-by, as it is on a bend near the summit of the road). Park just off the road at the entrance to the disused army camp at [NG 8480 7885].

Locality 5.10b Loch Tollaidh (Tollie)
From the car park, the Tollie antiform is clearly visible in the low cliffs on the south side of the loch: the foliation is vertical at the east end, then becomes flat in the centre and suddenly becomes vertical again at the west end. Cross the road and join the track, ford the Tollie Burn, then go through a gate in the deer fence and walk to the exposures. At [NG 8505 7847] the first exposures show a vertical foliation in the gneisses, deformed by isoclinal folds and small z-shaped folds. A horizontal foliation is also easily seen, as well as a lineation. Here there is also a vertical foliated amphibolite

dyke that clearly cuts across this lineation and which contains small-scale folds and shear zones, indicating that the dyke was deformed and metamorphosed after it had intruded the already-folded gneiss. Following the exposures to the southwest, note that the gneiss and amphibolite become horizontal over the antiform.

The Tollie antiform and the Gairloch shear zone are related as a pair of structures that Park (2010) describes as a shear-fold, and this is one of the most important and complex structures in this part of the Lewisian outcrop. The hinge-line of the Tollie antiform marks the northeastern margin of the 6.5km-wide Gairloch shear zone.

Return to the vehicle and drive 2km to the west end of Loch Tollie and park in the old quarry at [NG 8300 7817].

Locality 5.10c Tollie quarry

The rocks at the back of the quarry are in the steep limb of the Tollie antiform. Coarse black and white striped gneiss, a mixture of amphibolite and quartz–feldspar gneiss, shows good examples of small-scale folds with axial planes parallel to the near-vertical foliation. Some 15cm-wide quartz veins cut the foliation and are themselves flattened and sheared. If the amphibolite was originally a dyke, it is now perfectly parallel to the foliation in the gneiss and no cross-cutting relationship can be seen, so the amount of strain is very high. This point marks the edge of the Gairloch shear zone.

Continue towards Gairloch. The red cliffs on the right around Lochan nam Breac, opposite a forest plantation, are flat Torridonian sandstones, occupying a hollow in the Lewisian landscape [NG 8151 7797]. The disused roadstone quarry at this point is in vertical Lewisian rocks. Stop briefly in the large lay-by overlooking Gairloch [NG 8045 7745]. The road cutting is in vertical black Lewisian hornblende schist. Note the contrast in the landscape between the knock and lochan formation on the gneiss and the much gentler slopes on Torridonian rocks to the west. For the best views of the Torridon and Applecross hills to the south, drive west for 4km along the B8021 towards the youth hostel at Carn Dearg [NG 7650 7750].

Drive south of Gairloch for 2km and park at the end of the pier road [NG 8080 7507]. On the way, note the Torridonian rocks exposed on the beach below Gairloch Hotel, which is sitting on a raised beach cliff.

Locality 5.11 Gairloch pier

The rocks exposed at the pier [NG 8080 7507] are in the Ard gneiss, named after the headland that forms the southern edge of Gairloch beach. It is a coarse grey rock, rather massive, cut by many pink quartz–feldspar veins, and the foliation is vertical when it can be seen. It has a very distinctive texture, containing granular 3mm feldspar crystals that have been flattened into augen (eye-shaped). The Ard gneiss has been dated at 1900 million years, and is a deformed sheet of granite or granodiorite that was intruded into the metasedimentary schists of the Loch Maree Group during Laxfordian events. From the chemical composition of the gneiss, Park *et al.* (2001) concluded that it probably formed from partial melting of oceanic material before collision with continental crust. Follow the marked track from here over the headland towards the beach. At the top of the hill the gneiss varies in colour from brick red to pink and is mostly very coarse and quite homogeneous. Wherever the foliation can be seen, it is folded and the folds are cut by 20cm-wide quartz veins [NG 8029 7517]. Continue on the track towards the small promontory of An Dùn. At the foot there is a clear contact between the Ard gneiss and a black, finely foliated and folded amphibolite sheet containing narrow folded veins. It also contains blocks or xenoliths of folded gneiss [NG 8031 7534] and must therefore be younger than the gneiss. Go down onto the rocks at sea level below An Dùn and note how the pattern of two intersecting cleavages produces diamond shapes. Glacial whalebacks at sea level are well expressed here. Walk along the beach to the north end for a magnificent example of the Lewisian–Torridonian unconformity [NG 8058 7580]. Here the gneiss blocks are plastered against red sandstone in a highly irregular fashion, indicating the piling up of scree on an exposed bluff of gneiss (Fig. 5.11).

Take the track that leads from the beach to the car park opposite the church and walk down the main road back towards the pier. The next locality can be done on foot. Turn left into the car park beside a pond adjacent to the Old Inn [NG 8095 7502].

Locality 5.12a Flowerdale – Loch Maree Group

The high crags on the hills around here are made of amphibolite, and the valley has been eroded into softer schists and marbles. Cross the footbridge, go behind the inn, past a DIY store on the left, through a wooden gate onto the footpath through the trees, and up a slight rise. Finely laminated,

Figure 5.11 Basal breccia of Torridonian on Lewisian at Gairloch beach.

vertical crenulated (crumpled) schists with a very pronounced lineation, rusty brown on the weathered surface, grey when fresh, form a rocky knoll on the right above the second gate at [NG 8150 7486] by a stile. The schists are part of the Loch Maree Group of metasedimentary rocks. Here the schists are vertical and highly deformed, and lie within the Flowerdale shear zone, one of four belts in the much wider Gairloch shear zone (see Park, 2010 for detailed maps and a description of the structural history). Take the left fork on the path 25m after crossing the stile; the vertical green rock on the path at the fork is the Kerrysdale amphibolite. Go onto the wooded ridge to the right before a footbridge across the stream at [NG 8187 7492]. The blue marked path follows this ridge, which is made of vertical, finely laminated grey, black, brown and bluish schists, quartzite and marble. Some of the quartzites contain magnetite and are known as banded ironstones; a compass needle will be deflected when held against the rock. It is the presence of abundant iron that gives the schists their rusty weathering, due to oxidation of the iron. Careful examination will show many examples of minor folds in the schists. There are several small ridges and hollows hereabouts, owing to the different hardnesses of the various types of metasediment. One such ridge, above a paddock at [NG 8183 7483], has exposures of the ironstone. Follow the marked (red) path to Flowerdale falls, noting the vertical black schists in the riverbank at a second wooden bridge [NG 8223 7491]. Much of the ground is overgrown with long grass, bracken and brambles. Take the path past the farm and Flowerdale House back to the main road. For more details of the Loch Maree Group and the Lewisian geology of Gairloch, see Park (2002, 2010), and an excursion by Park in Barber *et al.* (1978). The description of the Loch Maree Group as a subduction–accretion complex has been referred to previously; see Park *et al.* (2001) for the field evidence, and a detailed explanation.

From Gairloch drive south, then east through Kerrysdale gorge with its dam and hydroelectric scheme. On the way up the narrow gorge, note the vertical black and rusty-brown schists of the Loch Maree Group on the road cuttings at the steepest part of the hill, just before the dam is reached. Copper, zinc and gold mineralization in these 2000 million year old rocks has been interpreted as the result of hot, ore-rich volcanic fluids being injected into sediment on new sea floor caused by continental rifting.

Locality 5.12b Kerrysdale dam

Park at the dam and cross the road, then follow the path through the forest until a clearing is reached at the top of a hill. The rocks seen on and from the footpath are amphibolites, which were originally basalt lavas or sills. Bright yellow, orange and brown folded schists and quartzites are seen at ground level in the clearing [NG 8376 7250]. The rusty colour is due to the weathering of iron and copper sulphides in this ore deposit, termed a gossan. Rather fresher sulphide crystals – bright yellow iron pyrite and duller, brassy orange-yellow chalcopyrite (copper sulphide) – can be seen in the fold hinges (Fig. 5.12). Lead is also present, and rare gold. Mineral exploration in the late 1980s did not result in any commercial extraction of gold. Return to the road and continue east to Loch Bad an Sgalaig [NG 8500 7199] and park in the lay-by.

Locality 5.13a Loch Bad an Sgalaig

Climb up to the small round rocky hill, Meall an Tuill-aoil (Gaelic: the hill above the limestone hollow) opposite the lay-by. This is part of a very thick, vertical dark blue to dark green foliated amphibolite that was intruded into the metasedimentary schists as a gabbro, then deformed and metamorphosed to amphibolite. It runs northwest–southeast and can be traced for a long distance, making a prominent landscape feature. Continue

Figure 5.12 Mineralized Loch Maree Group metasediments near Flowerdale, showing altered rusty ore-rich pyrite 'gossan'.

east for another 700m and park on the left at the fishing hut opposite Am Feur Loch [NG 8568 7210].

Locality 5.13b Am Feur Loch

Walk 100m up the grassy slope on the hill above the hut to a small hollow [NG 8555 7221] by a bright green grass- and bracken-covered knoll that marks an old limestone quarry. Beware of the plantation hollows, which are often hidden by bracken and heather, and which may be full of water. On the weathered surface this marble is khaki brown, with a gnarled appearance and vertical dip. It is creamy and speckled white on fresh faces, and is cut by thin veinlets of white calcite. Surrounding the marble is pale pink and grey, highly sheared gneiss with a vertical, closely spaced foliation. This rock is the Am Feur Loch marble, one of several in the Loch Maree Group (see Park, 2002, for a detailed description and map of the Lewisian geology of Gairloch and Tollie, also an excursion by Park in Barber *et al.*, 1978).

Return to Gairloch for accommodation, or continue to Loch Maree on the A832 road (Chapter 6).

Chapter 6

Gairloch to Kyle of Lochalsh

Introduction

This chapter, covering Wester Ross, includes the type area for the Torridon Group and the Diabaig and Applecross formations. Precambrian palaeo-relief on Slioch is magnificently exposed, and the upside-down stratigraphy caused by folding and thrusting at the edge of the Caledonian mountain belt is a classic feature. Superb examples of glacial erosion and deposition are seen – corries, moraines, and ice-moulded landforms. Wester Ross is a National Scenic Area, while Torridon is cared for by the National Trust for Scotland.

Locality 6.1 Loch Maree; Slioch

Drive south then east from Gairloch on the A832, crossing Lewisian gneiss exposures (Fig. 6.1). Stop in the large car park at Talladale, overlooking Loch Maree [NG 9376 7036]. Rock slabs at the car park are coarse red Torridonian sandstone. Opposite is the imposing mass of Slioch, made of horizontal Torridonian beds covering hills and valleys of rough and knobbly Lewisian gneiss, and marking one of the most impressive unconformities in the country (Fig. 6.2). At the time the Torridonian sediments were being deposited by large rivers, these hills of Lewisian rocks were exposed at the surface, then blanketed by sandstone so that the hollows and valleys were filled in and the sandstone eventually piled up and covered the hills. These are now in the process of being exposed again, to produce an exhumed landscape. Note the very straight northern shore of Loch Maree, which lies within a large fault, excavated by ice. This NW–SE fault is Precambrian in age, much older than the NE–SW Caledonian faults that tend to dominate northern Scotland.

Drive southeast for 10km and park on the right at the information centre for the SNH Beinn Eighe National Nature Reserve (established in

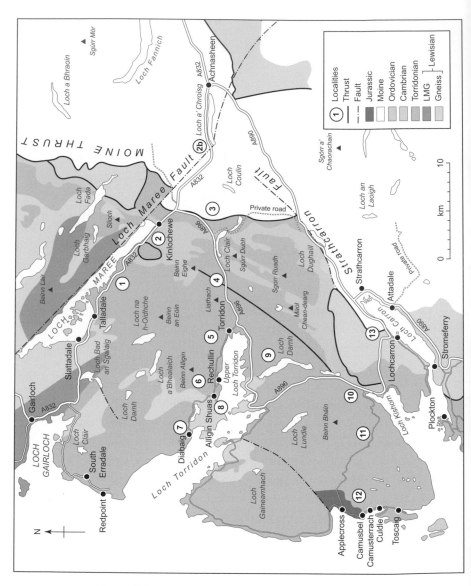

Figure 6.1 Route map with geology: Gairloch to Kyle of Lochalsh.

Figure 6.2 Loch Maree; Slioch, showing Torridonian overlying hills and valleys of Lewisian, forming a fossil landform.

1951 as the first in Britain). There is an illustrated leaflet for the mountain walk (as well as one for the woodland trail, plus geology display boards).

Locality 6.2a Beinn Eighe

Follow the mountain trail, which starts under the bridge, and go through the forest path, on Torridonian rocks – sandstone and conglomerate with small, white well-rounded quartzite pebbles. In places glacial striae can be seen on the slabs. Once the path suddenly begins to narrow and climb steeply, the rock exposures are of white Cambrian Quartzite. This is a coarse gritty rock in which individual white quartz grains are very obvious. The path follows the dip slope, and bedding takes on a steeper dip above the tree line. Here, the quartzite is highly shattered, being close to a fault, and angular chips on the path derive from quartzite scree. Note the white cliffs higher up Meall a' Ghiubhais, where the bedding is very obvious. Good examples of cross bedding in the quartzite can be seen in the exposures on the path at the 'Geology' signpost [NH 0003 6411].

Take the steps that go up higher onto wide bedding planes (a fossil beach in effect), covered in a variety of the Pipe Rock known as the Trumpet Rock, in which the worm tubes are 5cm in diameter (Fig. 6.3). Smaller

Figure 6.3 Beinn Eighe: bedding plane of Cambrian quartzite (Trumpet Rock), showing circular impressions of worm tubes.

pipes are seen on the path lower down. Glacial striae running in two directions can be seen on the white slabs at this locality. From this point there is a clear view across Loch Maree to Slioch on the left and Beinn a' Mhùnidh on the right. Between the two hills is the deeply eroded valley of Gleann Bianasdail, excavated along a northeast–southwest fault at right angles to the Loch Maree Fault. On the west side of the fault is dark Lewisian, while to the east (the downthrown side) the white Cambrian rocks lie unconformably on top of dark brown horizontal Torridonian sandstones, then to the east towards the top of the hill the knobbly Lewisian can be seen again, this time thrust over Cambrian and Torridonian by the Kinlochewe Thrust. Gleann Bianasdail is a hanging valley, left above the level of the main glacial trough of Loch Maree. Continue on the footpath to the viewpoint cairn at the top of the hill at 550m [NG 9930 6330], still on Pipe Rock. From here, Beinn Eighe is the dominant mountain, and there are distant views to Ben Wyvis in the east and the mountains of Kintail to the south.

Plant life on the plateau is similar to Arctic tundra. Glacial erratics of Torridonian sandstone are found at this level. Note also the distinctive ice-moulded topography around here, where the smooth, rounded ridges of white quartzite are strongly aligned in parallel formation. The hill above to the left (west) with its distinctive shape is Meall a' Ghiubhais (887m),

Figure 6.4 Beinn Eighe mountain trail: folded Torridonian mylonite at thrust plane.

which is Torridonian, thrust over Cambrian by the Kinlochewe Thrust. A thrust-bound outlier such as this is known as a klippe (German: cliff). Skirt the small lochans and head north, following the path on the way down. The deep cleft with a waterfall at the foot, seen from [NG 9898 6441], is a north–south fault. At [NG 9913 6462] the bright green vegetation marks an exposure of the Fucoid Beds, shaly beds rich in lime and potash, producing more fertile soil than the acid quartzite. Below this point the path re-enters the quartzite, which again shows glacial striae in a number of places. At the small waterfall at [NG 9898 6453], the green, folded striped rock on the stepped path is mylonitized Torridonian, within the thrust zone (Fig. 6.4). Return to the bottom of the track and then to the vehicle.

Drive 3km southeast to Kinlochewe (accommodation, toilets, shops, petrol, post office). From here, if there is time, it is worthwhile making a detour to Achnasheen, 15km east along the A832 road, to see the remarkable glacial lake terraces. Stop briefly at the viewpoint [NH 0663 5930] looking down Glen Docherty, which is a splendid example of a U-shaped valley excavated along the Loch Maree Fault. Note how the moraine hummocks on the sides of the valley have been cut through by the modern stream, exposing the irregular nature of the sand, gravel and boulder material. The hills on either side of the valley are Moine schists. At Achnasheen, turn right onto the A890 and go south at the roundabout for about 500m, then park near Ledgowan Lodge. At Achnasheen Station there is a display board explaining the glacial lake phenomenon.

Locality 6.2b Achnasheen (optional detour)

Looking east from here, there are excellent views to the extensive flat terraces formed by deposition of sediment in a glacial lake that existed during the deglaciation stage in Strath Bran when the valley was dammed by ice from the Fannich mountains and from glaciers in the west. The terraces mark six levels of the lake, the highest being 190m above sea level and 40m above the present valley floor. In addition there are also a number of deltas that formed when sediment was laid down by meltwater streams flowing into the lake (Fig. 6.5). This locality is the best known and most important example in the country of glacial outwash terraces, and has been described as a classic of its type (Gordon & Sutherland, 1993). Peach and Horne (of Assynt fame) were the first to describe this locality and to draw the correct conclusion about its glacial meltwater origin.

Return to Kinlochewe and turn left onto the A896, towards Torridon. Stop after 5km, opposite the track that leads to Loch Clair [NH 0020 5815].

Locality 6.3 Beinn Eighe, Liathach, Loch Clair

Park at the roadside then walk across onto the footpath that crosses the bridge and walk 500m down to the shores of Loch Clair for an excellent

Figure 6.5 Achnasheen: sediment-filled post-glacial lake.

Figure 6.6 Liathach from Loch Clair; Torridonian

Figure 6.7 Beinn Eighe from Loch Clair; Cambrian (white quartzite screes) on Torridonian unconformity.

view west to Beinn Eighe and Liathach. The unconformity on Beinn Eighe between Cambrian Quartzite with its white scree slopes on top of horizontal brown Torridonian sandstone is very clear, as is the bedding on Liathach (Figs 6.6, 6.7). The two rock groups have been piled up by a series of small thrusts, so that they are repeated several times as slices within the Beinn Eighe imbricate stack. For a detailed and very advanced account of the complex thrust geology of this area, see Butler *et al.* (2007). This study is complemented by an excellent website, containing many photographs, explanatory diagrams and maps (see Appendix for details). The prominent large hill west of Loch Clair is Sgùrr Dubh (782m), which has Cambrian on top of Torridonian, interleaved by folding and thrusting, as on Beinn Eighe. Westerly directed thrust movements have caused the rocks to be pushed up into much steeper angles, up to 60°. Pre-existing faults, related to the Loch Maree Fault, and original irregularities in the Lewisian–Torridonian boundary may have been responsible for the development of the complex fold and thrust structures around Beinn Eighe (Cain *et al.*, 2016).

Along the Loch Torridon road, note the undulating nature of Torridonian on Lewisian rocks as the road goes along the unconformity surface, which at the time of Torridonian deposition would have been equally undulating. Stop at the large car park on the right in Glen Torridon [NG 9568 5680].

Locality 6.4 Glen Torridon

Opposite the car park is the entrance to the remarkable Valley of a Hundred Hills (Coire a' Cheud-chnoic). These are moraine hummocks, made of unsorted boulders, sand and gravel, formed at the snout of a melting valley glacier at the end of the Loch Lomond Readvance, 10,000 years ago. Most are elongate in a north–south direction, but from the viewpoint they look conical. Elongation probably occurred by ice shaping the moraine that had been left in the valley by the melting of the earlier Devensian ice sheet that covered the entire Highlands from 20,000 years ago and melted 13,500 years ago. This classic locality is the best example of its kind in the country (Gordon & Sutherland, 1993). Rock exposures at the top of the ice-scoured hill (Sgùrr Dubh) above the moraine show several horizontal ledges of white Cambrian Quartzite, intensely thrust and interleaved with brown Torridonian sandstone (see locality 3 above). There are small white scree slopes beneath each quartzite ledge, showing the contrast very well (Fig. 6.8). This is a part of the Moine Thrust Zone known as the Achnashellach culmination (a bulge caused by the stacking up of thrust slices), similar to but on a smaller scale than the Assynt culmination farther north (Chapter 4, locality 10), and without the igneous intrusions. The Achall culmination at Ullapool has also been mentioned in Chapter 5. Watkins *et al.* (2015) describe the fold and thrust structures within this culmination in great detail, well illustrated with maps and cross-sections.

Carry on west through Glen Torridon, noting small hillocks on the valley floor and the huge round boulders. This is part of an end moraine and outwash fan related to melting of a local glacier at the end of the Loch

Figure 6.8 Valley of a Hundred Hills, Glen Torridon: hummocky moraine; above and to the left are horizontal thrust wedges, marked by white Cambrian quartzite screes.

Lomond Readvance. Where the interior of the hummocks is exposed, the wide range of sizes can be seen, with coarse sand predominating. About 2km east of the Youth Hostel [NG 9200 5568], flat surfaces at the roadside are in Torridonian rocks, with small flakes of fine pink mudstone and small smooth, rounded pebbles of quartzite in a sandstone matrix. The mudstone flakes indicate drying-out of the land surface, then transport by a river. Pass the foot of Liathach, and look up at the awe-inspiring gigantic staircase made of completely bare beds of Torridonian pebbly sandstone (Applecross Formation, 3km thick), here with a gentle dip to the east. At the top, the four highest peaks have a cap of Cambrian Quartzite, each with a skirting of white scree.

Take the minor road leading west past the Youth Hostel at Torridon, passing over the nearly flat Torridonian until the Lewisian is reached at Alligin Shuas, where the road turns north and becomes much more twisty. A northeast–southwest fault excavated by the River Alligin marks the contact between Lewisian and Torridonian here. Looking south, the unconformity can be seen very well in the landscape on the opposite side of Loch Torridon.

After going through the narrow gap of Bealach na Gaoithe (Gaelic: pass of the wind), stop on the hill of Sidhean Mòr at a small lay-by on the left, overlooking Loch Diabagas Airde (Upper Loch Diabaig), near where the road descends steeply [NG 8240 5963].

Locality 6.5 An Ruadh-mheallan

The prominent hill above the road, An Ruadh-mheallan (Gaelic: the red rounded hill), is made of Torridonian Applecross Formation. By the road are extensive ice-smoothed bare exposures of Lewisian rocks, mostly coarse, creamy white or pale grey homogeneous quartz–feldspar gneiss, sometimes banded, sometimes with no visible structure, dark ultrabasic layered igneous rocks, white anorthosite and migmatite, deformed by numerous small-scale folds and cut by many northwest-trending Scourie dykes. This area, together with the Scourie to Loch Laxford area (Chapter 3) is important historically, as it was used by Sutton and Watson in the early 1950s to illustrate their interpretation of Lewisian structure. They held the view that the dykes formed in a single episode of intrusion and that the rocks cut by the dykes were formed during the Scourian episode of folding and high-grade metamorphism, while structures such as shears, folds and lower grade metamorphism affecting the dykes (Scourie dykes) were defined as belonging to the younger Laxfordian set of events. In their opinion, the dykes could be used as time markers to separate folding and metamorphic events. More recent work has shown that there are several sets of dykes and that they were intruded into existing shear zones, inferred to have formed during the Inverian episode (or late Scourian, with early Scourian now referred to as Badcallian), and many were then sheared internally, and more especially at their margins, to produce a schistosity. See Wheeler (2007) for a discussion of this area, together with detailed maps. This particular locality is notable for having broad areas that are relatively unaffected by later movements (termed low-strain zones), which are concentrated in narrow, steep shear belts (high-strain zones) that surround these areas. Dykes within the shear zones are usually thin and foliated, with the foliation parallel to the sheared gneiss, whereas those in the homogeneous grey gneiss are often thicker and have a narrow foliation at the margin only, while the interior has a random igneous texture. Good examples of both cross-cutting and concordant dykes can be seen between the hill of Sidhean Mòr and Loch nan Tri-eileanan (Gaelic: loch of the three islands) about 500m east of the road.

Drive down the steep hill and stop after 1km opposite Loch Diabagas Airde at a bend in the road just beyond a small stream [NG 8190 6010].

Figure 6.9 Upper Diabaig: red siltstone and mudstone, showing ripple marks in section.

Locality 6.6 Upper Diabaig (Diabagas Airde)

The road cutting here shows finely laminated grey and pale reddish mudstone, shale and sandstone with ripple marks – in cross-section these produce a wavy pattern (Fig. 6.9). Actual bedding planes can be seen in the next locality. This is the type area for the Diabaig Formation, the lowest part of the Torridon Group. Just to the east there is an exposure of coarse breccia made of angular gneiss fragments, close to the Lewisian, representing the unconformity. Since the gneiss forms higher ground around here, the Diabaig shales must have been deposited in a hollow in the contemporary landscape.

Drive down to Lower Diabaig, passing the steep Lewisian hill of An Tòrr on the left. At the sharp bend there is a good view into Loch Diabaig, surrounded and almost enclosed on the south, east and west by Lewisian cliffs. Park near the pier at [NG 7976 5986].

Locality 6.7 Lower Diabaig

The pier is built on dark red Lewisian gneiss with pegmatite and migmatite. It is directly overlain by a sedimentary breccia made of angular blocks of gneiss and white quartzite – this is the basal unconformity with the first beds of the Diabaig Formation (see Table 2.2). Since the breccia is also red and the gneiss fragments are surrounded by fine quartz and feld-spar, it is rather difficult to see the actual contact. Walk south along the shoreline track for 400m to the edge of a cliff of Lewisian gneiss. Here the banding and patchy pegmatite and migmatite are folded and cut by a vertical Scourie dyke, which shows a strong foliation at the edges. Clearly the dyke is younger than the folding. Return to the pier, then walk north for about 100m along the road, then left on to a footpath past some boat sheds to avoid slippery boulders on the beach. Grey, fine-grained, laminated Diabaig shales form the exposures at sea level and in the small cliff beneath the trees here. Many examples of ripples and mud cracks at various scales can be seen on the bedding planes (Figs 6.10, 6.11), indicating shallow-water conditions and frequent drying out of mud at the surface. Two strong sets of joints at 60° break the shales into lozenge shapes. Small hard nodules (called concretions) rich in phosphate occur in some of the shale beds,

Figure 6.10 Lower Diabaig: ripple marks on mudstone, with possible algal markings.

Figure 6.11 Lower Diabaig: desiccation cracks on Diabaig mudstone.

Figure 6.12 Lower Diabaig: trace fossils possibly related to decaying algae, on Diabaig shale bedding plane.

and contain fine algal filaments and possible spore walls, making them the oldest fossils in Scotland. These are tiny microfossils, but they appear to be quite abundant and well preserved in the nodules. Other indications of biological activity on the bedding planes include sharp star-shaped reticulate patterns from algal mats, and tiny round blisters, possibly caused by gas bubbling up from decaying algae and bacteria (Fig. 6.12). See Brasier *et al.* (2017) and Wacey *et al.* (2017) for well-illustrated discussions of these features. Based on these, an age of 1000 million years has been given for the base of the Diabaig Formation. The sediments were most probably deposited in a shallow lake. On the hillside just west of here, the broad slabs are made of Applecross Formation, the next unit above these Diabaig beds. The Applecross sandstones were transported into the area by rivers coming from the west, i.e. Greenland, across a quartz–feldspar gneiss basement.

Locality 6.8 Upper Loch Torridon

Drive the 15km back to Torridon and turn right onto the A896 towards Shieldaig. From Annat [NG 8938 5445] there is a superb view of Liathach at the head of the loch, with Beinn Eighe in the distance. The road then goes along the south shore of Upper Loch Torridon, and the frequent bays are eroded into Diabaig shale, which is softer than both the Applecross sandstone and the Lewisian gneiss. At Òb Gorm Beag opposite Alligin the Torridonian–Lewisian unconformity is exposed. Alligin, Liathach and Beinn Dearg are well seen from the large car park viewpoint at [NG 8658 5421], with dark red, coarse Applecross sandstones making the large flat slabs in the car park. Glacial valleys between these mountains are very prominent, as are the alluvial scree apron and avalanche scars at Torridon House. An Ruadh-mheallan (locality 5) can be seen opposite. Note also the Lewisian gneiss on the lower peninsula out to the west at the end of Upper Loch Torridon, below Torridonian hills. Another good view of the change in landscape between Torridonian and Lewisian is obtained at the roadside by Balgy Lodge [NG 8453 5445]. Stop above the next large bay at [NG 8265 5383].

Locality 6.9 Òb Mheallaidh, Upper Loch Torridon

Òb Mheallaidh is the largest of the bays along the loch, and it is almost entirely enclosed. The back of the bay (bedding planes are exposed at the roadside by a small waterfall) is formed of greenish-grey rippled Diabaig silt-stone, laid down in a valley between two hills of gneiss. Adjacent road cuttings on the east side of the bay [NG 8316 5364] show the junction between the Lewisian gneiss and the Applecross Formation, which forms the terraced slopes on the high ground of Ben Shieldaig, above a Lewisian hill.

Continue south through Glen Shieldaig. Shieldaig Island (National Trust for Scotland) is completely covered in Scots Pine. Notice the pines growing on ledges of Torridonian on the hillsides above the road.

At Couldoran there is a sudden change from rough heather moorland to bright green grassy fields at the junction between Torridonian sandstone and Durness Limestone, with the mountains of Applecross rising up majestically to the west. The Rassal Ashwood National Nature Reserve, the largest of its kind in northern Scotland, is here because of the fertility of the soils on the limestone. The limestone, which weathers to a dark blue colour, shows some features of karst topography (clints and grykes), but not to the same extent as at Inchnadamph in Assynt (Chapter 4, locality 5). Copper was mined here in the mid-18th century from a vein in the limestone, still visible on the hillside above the woods [NG 849 432].

Turn right at Tornapress, 1km south of Rassal, onto the minor road to Applecross. Stop in the lay-by beside the rocks exposed on the west side of the bridge over the River Kishorn, An Drochaid Mhòr (Gaelic: the big bridge) [NG 8342 4230].

Locality 6.10 Loch Kishorn

Cuttings beside the road show typical Applecross Formation red and brown sandstones, with almost flat bedding that is disrupted by contortions caused by compaction and slumping of wet sediment at the time of deposition. Cross bedding is also very common here, caused by currents flowing in different directions. The characteristic red colour is due to the presence of a substantial amount of feldspar derived from underlying Lewisian gneiss, combined with hematite cement. Sandstone with this amount of feldspar is known as arkose, although the preferred term is now feldspathic sandstone. Looking due east from the bridge, the dark hill above the Tornapress road junction (Cearcall Dubh) exposes knobbly Lewisian gneiss on top of Torridonian terraces, which in turn lie above green fields and woods, with an outcrop of dark blue Durness Limestone at road level. Here the normal stratigraphy is completely upside down – the rocks have been folded and thrust above the Kishorn Thrust, which has exploited the weak Durness–Torridonian junction.

Continue west up the very narrow, steep and twisting road towards Applecross. This is a difficult climb, and not to every driver's taste on account of the hairpin bends and precipices, but it is exciting. The alternative coastal route starts 1km south of Shieldaig, but this is very long and geologically not so interesting. The first part of this road undulates between Lewisian and Torridonian. At Ardheslaig [NG 7804 5603] pebbles of gneiss and quartzite can be seen in Torridonian conglomerate in a hollow eroded in Lewisian, while 9km to the northwest at Fernbeg [NG 7353 5966], Applecross sandstones show cross bedding and contorted bedding. This is the type area for the Applecross Formation. Beyond this point the road turns south and follows Torridonian bedding planes as far as Applecross village (locality 12), but rock exposure is generally poor.

Locality 6.11 Bealach nam Bà (pass of the cattle)

Stop at the viewpoint to the west of the pass [NG 7745 4258], to admire the glacial erosion features where the red Torridonian has been deeply carved by the ice, and the landscape to the west. The Palaeogene volcanic mountains of Skye can be seen, as well as Dùn Caan on Raasay (basalt lava on top of Jurassic sedimentary rocks) and the island of Rona (Lewisian gneiss). Continue to Applecross village and turn left (south) towards the pier [NG 7081 4460].

Locality 6.12 Applecross

Broad, gently dipping bedding planes on the shore above high tide mark halfway between the last house and the jetty (beyond the main pier) show grey limestone with weathered-out hollows on the surface and a rectangular pattern of joints, filled with grass. There are broken fragments of fossil brachiopods and corals, which indicate that the rocks are Jurassic in age. At the back of the beach, a green sandstone occurs beneath the limestone, as well as pebble beds with 1cm round white and pink quartzite pebbles in a coarse cream-coloured sandstone. Jurassic rocks are rare on the west coast of the mainland, but thick sequences have been preserved beneath the Palaeogene lavas on Skye. The beach here has a number of large boulders of Torridonian sandstone on top of the limestone bedding planes; these have been derived locally, and probably transported by ice (Fig. 6.13). Such blocks are known as erratics. From here there are views west to Skye, Raasay and Rona.

Return to Kishorn and turn right onto the A896, following the outcrop of Durness Limestone in the valley (bluish grey cliffs, rich green vegetation)

Figure 6.13 Applecross: Jurassic limestone pavement with rectangular joints, and rounded Torridonian erratic boulder sitting on top

Figure 6.14 View of Torridon hills from Kishorn.

as far as Sanachan, then east to Loch Carron. The Torridon hills from Kishorn make a remarkable sight (Fig. 6.14). Upside down Torridonian has been thrust over the limestone at this point by the Kishorn Thrust. Stop 4km east of Sanachan at some fresh road cuttings near the top of the valley of Abhainn Cumhang a' Ghlinne [NG 8739 4059]. The rocks here are Lewisian gneiss – coarse-grained dark red quartz–feldspar bands inter-leaved with black and dark green amphibolite and dipping at 30–40° to the east. Randomly scattered throughout the gneiss are 2mm white feldspar crystals that formed in a late overprinting metamorphic event. Banding in the Lewisian can be seen in cliffs opposite, on the north side of the road, with several small waterfalls. The rocks are strongly sheared and show slick-ensides (grooves) on fault planes. Irregular veins and patches of quartz up to 50cm across cut through the gneiss. Epidote is very common as distinctive cross-cutting bright green veins, especially in the amphibolite layers and along shear planes, oblique to the banding. As at locality 10, the rock sequence is inverted here in the thrust zone, i.e. Lewisian lies above Torridonian.

Continue east to Lochcarron, stopping at the corner of the road where it reaches the bottom of the valley, in a lay-by opposite a rock cutting near the village name sign [NG 8958 3952].

Locality 6.13 Lochcarron

The gently dipping black, dark green and purplish slaty-looking rock is mylonite derived from Moine schist, hence this point is the location of the Moine Thrust outcrop. Some open folds are present, also related to thrusting. The green sheen results from the presence of chlorite, a low-temperature metamorphic mineral. Further outcrops of folded mylonite can be seen on the southern end of the small peninsula of Slumbay Island 1km south of Lochcarron [NG 8964 3842], approached from a minor road.

Continue on the A896 through Lochcarron (petrol, shops, toilets, accommodation) and turn right onto the A890 at the head of the loch, passing the Strathcarron railway station. In Strath Carron, the road follows the line of the Moine Thrust. To the north the mountains of Sgòrr Ruadh and Beinn Liath Mhòr are made of Torridonian, repeatedly sliced together with Cambrian Quartzite and stacked up in a thrust zone. The road passes Moine schist as far as Attadale, then climbs the hill onto Lewisian gneiss. Along the south shore of Loch Carron, the road and railway share a very narrow strip of land. Safety netting at the entrance to the avalanche shelter now obscures a junction between Lewisian and Moine rocks, where strongly flattened and stretched gneiss pebbles in the Moine schist are taken to indicate that this is actually a sheared unconformity. Carry on south past Stromeferry to meet the A87 and turn right to Kyle of Lochalsh for accommodation (petrol, shops, post office, telephones, toilets, railway station). On the way down the hill at Auchtertyre (Lewisian; see Chapter 7, locality 3) there is a large lay-by with a good view of the Five Sisters of Kintail (Moine schist) and the Skye Bridge. Alternatively, take the narrow road to the small village of Plockton (National Trust for Scotland). Here, Torridonian rocks are found at sea level, but the higher ground is made of Lewisian gneiss, with its typical knobbly appearance, which has been inverted and thrust over the Torridonian. Very clear views are obtained from the hill of Rubha Mòr [NG 8068 3418], which can be reached on foot from the end of the row of cottages. A mild microclimate is responsible for the lush vegetation here, including palm trees.

Chapter 7

Kyle of Lochalsh, Glenelg and Glen Roy

Introduction

The geology in the area from Kyle of Lochalsh to Fort William is very varied, and includes Lewisian inliers within the Caledonian fold belt, and sections across the southern part of the Moine Thrust Zone, with interfolded Lewisian, Torridonian and Durness Limestone from the foreland, overthrust by Moine rocks, which are well exposed to the east. The Lewisian contains the only British examples of rare eclogites, which here are also unusually old. The Great Glen Fault is crossed at Laggan. In addition there are Caledonian granites, and the world-famous Parallel Roads of Glen Roy, dating from the end of the most recent glaciation. Landscape features include outstanding views of the rugged Five Sisters of Kintail ridge in Moine rocks.

The classic Glenelg–Attadale inlier is the largest Lewisian inlier east of the Moine Thrust. It consists of a western unit of grey and pink gneisses and garnet amphibolite of igneous origin, separated by a narrow shear zone of intense deformation containing fine-grained mylonites (the Barnshill shear zone) from a much more varied eastern unit of amphibolite, eclogite and marble, quartzite, ironstone and mica schist of sedimentary origin. There have been unsuccessful trials for copper and gold mineralization in the marbles and schists. The only other major examples of metasediments are at Loch Maree and Gairloch (Chapter 5). Eclogites are unusual high-pressure garnet–pyroxene metamorphic rocks, derived originally from basaltic material that was taken down in a wedge along a subduction zone 70km into the mantle. The Glenelg eclogites were formed 1100 million years ago in the Grenville orogeny, then brought up along shear zones into the middle crust, much closer to the surface. During this episode, 100 million years later, the rocks were altered from eclogite to amphibolite by the introduction of water along shear planes and crystal cleavages – a process referred to as retrograde metamorphism. Such old eclogites are exceptionally rare; they are not found in the main outcrops of the Lewisian on the mainland or the Hebrides. They were first identified and described in 1891 by Jethro Teall, of Scourie dyke fame (Chapter 3, locality 14). Later work was carried out by Peach and Horne, Sutton and Watson and John Ramsay, and more recently by Craig Storey (2008), and revisited by Ramsay and colleagues (Krabbendam *et al.*, 2018; see also the Moine guide for additional excursions and references, Strachan *et al.*, 2010). The structure is highly complex, involving several episodes of folding, thrusting and interleaving of Lewisian and Morar Group Moine rocks.

Locality 7.1 Kyle of Lochalsh

Park beside the disused Skye Bridge tollhouse and walk up the small hill – the Plock of Kyle [NG 7559 2738] (Fig. 7.1). From here there are

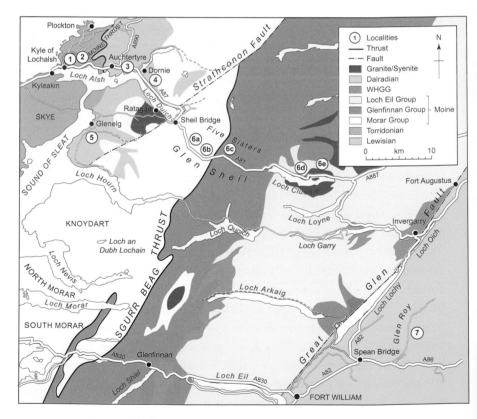

Figure 7.1 Route map with geology: Kyle of Lochalsh to Fort William.

Figure 7.2 Kyle of Lochalsh: horizontal quartz-filled fissures in vertical Moine schist.

views west to the Palaeocene igneous complex of Skye, and north to the Torridonian hills of Applecross (Chapter 6). High road cuttings opposite the superstore are of steeply dipping slabs of dark grey slates. The face is cut by a series of tension gashes arranged ladder-like (en échelon) and filled by white quartz (Fig. 7.2). The rocks are greenish-grey as a result of very low grade metamorphism in the Kintail Thrust zone, causing red iron (oxidized) to be reduced. About 1km east of the railway station, the roadside cliffs opposite the viewpoint at [NG 7720 2730] are overturned siltstone beds. At the boundary between grey and slightly reddish units is a soft, deeply weathered yellow band that may be a thin igneous sheet intruded into the siltstones (Batchelor, 2008).

Continue east on the A87 towards Balmacara, passing the monument to Donal Murchison, erected in 1863 by his great-grandnephew, the famous geologist Sir Roderick Murchison [NG 7871 2711]. Turn left at Balmacara and take the narrow minor road north towards Plockton.

Locality 7.2 Balmacara

Exposures along the road are of folded Torridonian sandstones with quartz-filled fracture zones. Cross bedding is everywhere upside down, showing that the rocks have been overturned. At a small quarry for the waterworks near the road junction [NG 7940 2949], the Torridonian is flat, strongly

Figure 7.3 Five Sisters of Kintail from Pass of Ratagan on the road to Glenelg.

jointed, sheared, and cut by many quartz veins. In the vicinity there are also a number of vertical bright red, shiny fault surfaces, coated in blood-red hematite (iron oxide).

From Balmacara, continue on the A87 and take the junction on the left uphill on the A890 in the direction of Stromeferry. Park in the large lay-by on the corner [NG 8490 2761], with views to the Five Sisters of Kintail (Fig. 7.3) and to Skye.

Locality 7.3 Auchtertyre (Western Lewisian)

Walk up the hill and examine the obvious shiny, bright green low exposure on the left of the roadside. These paper-thin rocks are rich in chlorite, and represent highly deformed and metamorphosed ultrabasic igneous rocks of the Glenelg Lewisian inlier, caught up in the Moine Thrust Zone. The large (2cm across) light green crystals are diopside, a calcium-rich pyroxene. There is an overall soapy feel to the rocks, caused by silvery-white talc – an alteration product of original olivine. Carry on up the road and cross over to the cuttings on the right-hand side. Here there is typical Lewisian gneiss: interbanded red, pink and black, with patches of dark green ultrabasic rock. In the lower parts of the exposure, the gneiss is very thin and streaky, where it has been thrust over the chlorite schists just seen lower down the road.

Return to the car park and drive south again, join the A87 and go 5km east towards Dornie. Park at Eilean Donan Castle [NG 8830 2591] (restaurant, toilets, tourist information).

Locality 7.4 Loch Duich

The exposures along the shore of Loch Duich present sections through the western and eastern Lewisian units of the Glenelg–Attadale inlier, east of the Moine Thrust.

Locality 7.4a Dornie

Low exposures of the western Lewisian at the Eilean Donan Castle visitor centre are in gently dipping, very dark basic gneiss, coarse, speckled and streaky, with hornblende, biotite and garnet, cut by pegmatite veins that are themselves deformed and disrupted, flattened and folded. Roadside exposures opposite the castle (take extreme care when crossing) show platy mylonite, crenulated (tiny folds like corrugations), with stretching lineations parallel to fold axes, indicating that the fold axes were gradually drawn out into the shear direction. Much of the Glenelg Lewisian resembles an immensely thick shear zone that probably formed during the Grenville orogeny, 1100–1000 million years ago. The original age of the Lewisian gneiss, metamorphosed in the granulite facies, is over 2000 million years. The heavily restored castle is built directly onto hornblende–biotite gneiss of the western unit.

Locality 7.4b Eilean Donan Castle viewpoint

Drive south for 500m and park in the lay-by at a prominent rock on the loch side, from where there is a view to the castle [NG 8855 2543]. Beware of traffic here, as Eilean Donan Castle has a seemingly mesmerizing effect on drivers. On the north side of the road, note the large fold deforming a pegmatite that cuts the gneiss; it has a fabric with the same trend as in the host rock. Amphibolite is also present, with a streaky network of white feldspar rock, and garnet-rich bands of granulite containing bluish quartz – a feature that indicates high pressure and temperature granulite facies metamorphism. The distinctive pale green lenses are of the mineral actinolite, an amphibole (as is hornblende) in actinolite–chlorite–talc schist, which is a metamorphosed ultrabasic igneous rock (compare locality 7.3). Additional roadside exposures are described in the Moine guide (Strachan et al., 2010).

Continue south and take the first road on the left after about 1km to Carr Brae (signposted) onto a narrow single track road.

Locality 7.4c Carr Brae (Eastern Lewisian)

Drive to the highest point on the road and park in the lay-by beside a viewpoint at [NG 8961 2462], looking towards Loch Duich and Skye. Cross the road and walk uphill to the foot of the cliffs of Creag Reidh Raineach. On the way, note the bright green grassy slopes, underlain by marble (metalimestone). Scattered widely over the surface are boulders of thinly foliated and strongly lineated pale grey and pink Lewisian gneiss, rich in mica and garnet, pegmatite and granite. Follow the track above the sheep fold to [NG 8974 2509]. Just beyond some grey mica schists are vertical exposures of khaki-brown marble, cream when fresh, heavily veined with white calcite, and very coarse-grained, showing a crude banding. Many small, round creamy nodules protrude from the rough surface. When broken open, these have large pale green, irregular feathery laths of the mineral diopside (a calcium silicate of the pyroxene family, found in metamorphosed limestone). Black amphibolite and hornblende gneiss, covered in heather, occur up the hill towards the crags, just beyond the marble. Below the cliffs [NG 8979 2544] there is sheared, crenulated and lineated mylonite and slabs of thin pink, white, green and grey flaggy gneiss in which the stripes or bands are perfectly parallel. These rocks are the product of intense flattening deformation.

If the weather deteriorates or visibility is poor, return to the road from this point, as the next part is somewhat remote, over rough ground. Otherwise, continue north round the end of the cliffs, then uphill to the right. Asymmetric folds are common in striped garnet amphibolite with hornblende–garnet pods at the core and later hornblende growth on the outside. Crenulation (crinkling) inside these pods results from deformation of an earlier metamorphic fabric. Coarse black biotite schists with redbrown garnets 2cm across occur at a cleft in the cliffs [NG 8982 2568]. Metre-scale folds are well seen here on the vertical faces, with a very strong rodding lineation parallel to the fold axes. Thin, platy quartz–feldspar–garnet–biotite gneiss, extensively folded, is common, often with elliptical black ultrabasic pods in fold cores.

On the top of the plateau, walk east towards Loch na Faolaig [NG 9044 2566]. At the very northwestern end of the loch is a large patch of eclogite containing coarse, bright green diopside and bright red garnets surrounded by haloes. On the cliffs on the opposite side of the loch [NG 90469 2578] are some excellent examples of recumbent folds (i.e. with flat axial planes). These are Moine rocks in the core of a large fold: flesh pink, thinly banded and striped quartz-rich schists, with a strong rodding parallel to the fold axes. On the hill above, between Loch na Faolaig and Loch na Craoibhe-caoruinn [NG 9058 2575], hornblende-rich Lewisian gneiss lies above the Moine rocks. High ground west of Loch na Faolaig (towards Carr Brae) is Lewisian, dipping beneath Moine. The Lewisian here is dark green striped hornblende gneiss and obviously different from the Moine rocks (the same folds are not present). Topographically the rocks are different too: Lewisian forms less well-exposed hummocks and knolls, whereas the Moine slabs occur in bare light grey-coloured cliffs. Take the grassy slope uphill opposite Loch na Faolaig, then down the steep valley to pick up the footpath again and head for the sheepfold and the road, taking care to avoid the cliffs of Creag Reidh Raineach.

Drive round the head of Loch Duich and stop in the car park just before the road enters a vertical cutting in the distinctive bright orange-red cliffs, before Kintail Lodge [NG 9418 2000].

Locality 7.4d Shiel Bridge

Walk to the road cutting exposures, which are in the Ratagan granite (Fig. 7.4). This is an igneous complex consisting of a number of related rocks,

Figure 7.4 Shiel Bridge: Ratagan granite, showing steeply dipping cooling joints.

mainly monzonite, diorite and syenite. It has been dated as 425 million years old (the same age as Ballachulish and Rannoch granites) and was intruded into Lewisian and Moine basement rocks towards the end of the Caledonian orogeny. Feldspar and hornblende crystals in the rock are roughly aligned, to form a steep fabric. It appears to have been deformed by movement along the Strathconon Fault, belonging to the Great Glen system of faults. The granite is cut in places (not here) by mineral veins with iron, molybdenum, copper, lead, zinc and bismuth sulphides as well as gold and silver minerals.

Drive on for another 1km and park at Shiel Bridge petrol station [NG 9400 1863]. Low exposures at the entrance to the car park are of grey quartz-rich mica schist (Glenfinnan Group Moine) with a strong rodding lineation and open folds that refold small-scale asymmetric folds, producing interesting fold interference patterns. A later mineral lineation clearly cuts the rodding.

From here, take the minor narrow, steep and twisting road at Shiel Bridge to Ratagan and Glenelg, taking the left fork up the hill towards Glenelg.

Locality 7.5 Glenelg (Eastern Lewisian eclogites)

Stop briefly in the Ratagan pass (Bealach Ràtagain) viewpoint car park [NG 9041 1987], from where there are splendid views to the Five Sisters of Kintail (Fig. 7.3). The ridge is made of Morar Group Moine rocks. To the left, on the northeast side of Loch Duich (on Sgùrr an Airgid, Gaelic: silver peak), the line of the Sgùrr Beag Thrust can be seen as a deeply eroded steep linear feature. Adjacent to the viewpoint the distinctive orange-red Ratagan granite can be seen in road cuttings. From here it is another 8km down Glen More to Glenelg.

In Glenelg, park at the crossroads triangle marked for the Brochs and the ferry [NG 8212 1981], and take the footpath route up the stream of Allt Mòr Ghalltair. Walk uphill towards the prominent knoll of the ruined circular Pictish fort Am Bàghan Galltair [NG 8212 2074], then to the ridge above this [NG 8222 2079]. Here the rocks are dark green garnet amphibolite, part of the Lewisian outcrop within a fold core, surrounded by Moine schist. From here there are good views to the Palaeocene volcanic islands of Skye, Eigg and Rum in the west, and to the Applecross hills of Torridonian sandstone just to the north. At this locality, eclogite can be recognized by its attractive glassy lustre (Fig. 7.5). It has dark grass-green pyroxene with

Figure 7.5 Glenelg: Lewisian eclogite rock (green pyroxene and red garnet).

bright red garnet and quartz. There is no feldspar. Eclogite is commonly retrogressed by the influx of water-rich fluids in veins and along cleavage planes to amphibolite, i.e. the pyroxene is converted to hornblende.

The next part should be attempted only in clear weather, and when the ground is dry, for it is across rough ground with no paths, and high grass, bracken and bogs. Head for the obvious ridge at [NG 8325 2117]. As at the previous location, there is some fresh eclogite, but much of it has been retrogressed to amphibolite. Better examples can be seen on the main ridge, by crossing a low wall and heading for low crags, about 500m away between [NG 8372 2109] and [NG 8366 2101]. Here, there are good examples of Lewisian mylonite, green forsterite–diopside marble (very rough-weathering surface, with a coarse sugary texture, full of clots and mineral clusters), eclogite, and amphibolite with boudinaged pegmatite sheets, approximately horizontal and parallel to the banding. The eclogites have a streaky texture and are cut by very thin rodded quartz veins, which must have formed during intense stretching. Considerably more detail on the highly complex geology of this area is provided by Storey (2008), including a field guide with colour maps and a discussion of the various possible interpretations for the evolution of these rocks. See also the Moine guide for a more extensive excursion (Storey, in Strachan *et al.*, 2010).

Figure 7.6 Glenelg: Dun Telve broch, constructed of flat slabs of Lewisian gneiss and Moine schist.

To return to the road, head due south for 600m then skirt the east side of Cnoc Mòr (bracken-covered slopes with no clear path) towards Iomairaghradain and then right (westwards) along the road back to Glenelg.

While in the area, it is worth visiting the well-preserved Glenelg brochs (Historic Scotland), about 4km away in Gleann Beag (signposted), Dùn Troddan [NG 8340 1724] and Dùn Telve [NG 8290 1725]. These drystone, double-walled fortified towers with internal staircases were constructed in the Iron Age about 2000 years ago, and are made of flat slabs of local Lewisian marble, gneiss, schist, amphibolite and mylonite (Fig. 7.6). The brochs were partly plundered for the construction of nearby Bernera Barracks in Glenelg Bay (1723).

From the brochs, return to the shore at Bernera beach [NG 8083 2041]. Beside Beachhaven cottage is the junction between Lewisian and Moine rocks. Both units are sheared, the junction being a tectonic one. Moine rocks can be easily recognized by their pale grey colour and the presence of cross bedding, and a pronounced lineation. The Lewisian, on the other hand, is represented by sheared green amphibolite at the contact with the Moine (Fig. 7.7). An obvious narrow pink dyke of lamprophyre cuts the

Figure 7.7 Glenelg: Lewisian gneiss at boundary with Moine schist.

Moine quartzite – it has an irregular contact and the margin is discoloured, due to the chilling effect during intrusion. Ramsay (in Strachan *et al.*, 2010) has a detailed map of this locality.

Return to Shiel Bridge and turn right onto the A87 and drive south through Glen Shiel along the Old Military Road, in the shadow of the towering sharp ridge of the Five Sisters of Kintail. An alternative would be to take the small Glenelg–Kylerhea ferry to Skye, then either drive south to Armadale and cross to Mallaig and start Chapter 8. The Sleat Group (Torridonian) at Kylerhea ferry consists of grey slabs resembling slates (they are weakly metamorphosed), with a steep dip seawards.

From Glen Shiel to Loch Cluanie the road passes the Morar, Glenfinnan and Loch Eil groups of the Moine succession (see Table 2.4), the Steep belt and Flat belt west and east of the Loch Quoich line, the Sgùrr Beag Thrust, and the Cluanie granite.

Locality 7.6a Morar Group Moine

Stop in the lay-by at [NG 9943 1340]. These exposures are of steeply dipping thick quartz-rich beds of grey psammite and thinner pelite (mica schist) belonging to the Morar Group. The mountains to the south, particularly the impressive ridge of the Saddle, show this well, the serrated summits reflecting the differential erosion of hard psammite (quartz-rich) and weaker pelite (mica schist).

Locality 7.6b Sgùrr Beag Thrust

Stop at [NH 0102 1341]. On the corner of the road is feldspar–garnet–quartz–mica gneiss cut by small veins of granite, which are deformed and broken, making the rock look like an augen gneiss. Walk up the stream above the road bend. In the stream bed are vertical white marble bands with green streaks, followed by black hornblende rock (metamorphosed ultrabasic igneous material), similar to the Lewisian at Eilean Donan Castle, locality 3. This is the basement beneath Moine rocks, which can be seen in an adjacent stream beside the forest fence, at [NH 0087 1364]. This (Glenfinnan Group Moine) is a highly sheared psammite. It is vertical and is now a finely banded quartz-rich mylonite in a zone of very high strain. Granite sheets here are thinned and show many tight isoclinal folds. The Sgùrr Beag Thrust can also be seen in riverside exposures (accessible only when the river is low) of mylonite at [NH 0066 1342], where the Glenfinnan Group sheared augen gneiss is against the grey uniform Morar Group, with the thrust marking the junction. For a map of this section, see Strachan *et al.* (2010).

Locality 7.6c Glenfinnan Group Moine

Drive east to [NH 0267 1234], where there are exposures of Glenfinnan striped pelites. These are coarse rusty-weathering garnet–muscovite schists, forming vertical cliffs opposite the large lay-by. Foliation surfaces are steep, as are the numerous fold hinges, which form a very strong rodding lineation. Vertical dips in the alternating quartzites and schists are clearly seen on the hillside beneath the Cluanie Forest ridge.

Locality 7.6d Loch Eil Group Moine

Along the Loch Cluanie shore between [NH 1237 1044] and [NH 1238 1049] are exposures of the Loch Eil Group (youngest Moine): coarse grey granular (sugary) quartz–hornblende–biotite–feldspar psammite, with recumbent isoclinal folds, and small cross-cutting pegmatite and granite veins, some of which are deformed. Some of the best structures are seen on the clean exposures around [NH 1249 1035], including refolded folds, sedimentary structures, boudins and quartz veins (Figs 7.8, 7.9). The overall dip of the rocks is 60–70° to the west. Cross bedding can be found away from the fold hinges, and indicates that the rock sequence is inverted. A large amphibolite pod (hornblende–feldspar–garnet rock, originally a basic igneous rock) is seen at the water's edge, one of a number of boudins strung out in a line, cut by an irregular network of diffuse granitic veins. A series of small tension gashes filled with granite can be seen commonly – these are from the nearby Cluanie granite. The sequence of folding here is quite complex; see the Moine Guide for detailed maps and explanations (Strachan *et al.*, 2010). Drive east for another 5km.

Figure 7.8 Loch Cluanie: folds in Moine schists.

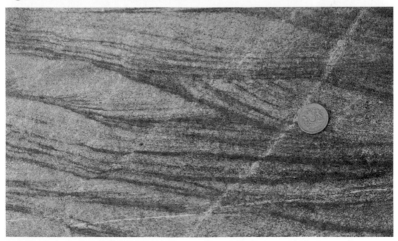

Figure 7.9 Loch Cluanie: cross bedding in Moine schists.

Locality 7.6e Cluanie granite

Stop in the lay-by beside the Cluanie dam, opposite the large quarry that was used for the construction of the spillway [NH 1773 1031]. To the south of the dam the round hill is Beinn Loinne in the Cluanie granite. Walk down to examine the exposures of the granite at the loch side, noting glacial striae on the smooth flat rock surfaces. The intrusion has been dated at 425 million years old and is undeformed, so was intruded after the Caledonian folding (i.e. it is post-tectonic). Veins of this granite cut the Moine rocks at locality 7.6d. It is a very coarse pink rock with large feldspar crystals (oligoclase) in which zoning can easily be seen, due to slight changes in composition as the crystals grew (Fig. 7.10). The dark crystals are hornblende and biotite. At this point there is a regional change in structural style in the Moine rocks: flat-lying foliation to the east, and steep to the west with many fold closures. This feature is referred to as the Loch Quoich line. Details of additional localities in this vicinity are to be found in the Moine Guide (Strachan *et al.*, 2010).

Drive east from Loch Cluanie and take the A87 towards Invergarry. If there is time, stop in the large lay-by above Loch Garry [NH 2114 0288] to view the contrast between the flat belt in the east, and the steep belt

Figure 7.10 Loch Cluanie granite, showing overgrowths on feldspar phenocrysts.

giving more rugged scenery in the west. The quarry opposite the viewpoint is a quartzite in the Loch Eil Group. Follow the A82 along Loch Lochy (in the Great Glen Fault) to Spean Bridge, then left on the A86 to Roybridge. Take the narrow minor road up Glen Roy (signposted from Roybridge; this road twists steeply uphill, and there are few passing places) and stop at the viewpoint with information boards [NN 2983 8533].

Locality 7.7   Parallel Roads of Glen Roy
This locality is in the Glen Roy National Nature Reserve and is a Site of Special Scientific Interest. From the viewpoint there is a superb panorama of the fossil shore lines, dating from the end of the most recent Ice Age, some 10,000 years ago, when the U-shaped valley was dammed by glaciers moving from the west during the Loch Lomond Readvance. Meltwaters filled the valley and the Parallel Roads indicate lake levels at 260, 325 and 350 metres (Fig. 7.11). These heights mark cols or drainage spillways on the surrounding hills, over which the lake waters drained into adjacent valleys. The shorelines were etched into the hillside at different periods as the glacier extended up the valley. In total, the lakes existed for 515 years. When the ice dam was breached at Spean Bridge, the waters of the lake

Figure 7.11 Glen Roy: Parallel Roads.

Figure 7.12 Glen Roy: recent landslips cutting across the Parallel Roads.

suddenly escaped beneath the ice and flowed rapidly out towards the Great Glen and into Loch Ness, then into the Beauly Firth. This type of catastrophic flood event is referred to as a jökulhlaup (Icelandic: glacier burst). Sand and gravel terraces at Fort Augustus and Inverness are sediments deposited by this event. Thomas Jamieson from Ellon in Aberdeenshire was the first to draw this conclusion in 1865. The gorge at Spean Bridge is a reflection of rapid down-cutting by fast-flowing meltwaters. It is possible to walk up the steep hill from the car park to the shorelines, which resemble flat tracks. Note the recent landslips which have cut across the lake shorelines in several places (Fig. 7.12).

The locality here is extremely important internationally for its contribution to the understanding of glacial phenomena. Various myths surrounded the origin of the Parallel Roads, and it was not until 1840, when the Swiss geologist Louis Agassiz visited the area, that the glacial origin was confirmed. Another famous visitor was Charles Darwin, who first published that they had a marine origin. He eventually accepted the conclusions of Agassiz in 1861.

If there is time, continue to the end of the road 6km to the north, for a view of the impressive outwash fan emerging from Glen Turret. This was deposited by meltwaters at the front of a valley glacier. Post-glacial river terraces are also remarkable here, reflecting progressive down-cutting as the land surface rebounded after the ice had melted. Other features in the glen include varve deposits – coarse and fine sediment layers deposited in summer and winter on the floor of the lake. For a comprehensive description and interpretation of Glen Roy and the surrounding area, see Sissons (2017). In the stream bed are exposures of steeply dipping Moine quartzite. Modern landslips are common in the glen, creating important landform features – several examples can be seen from the viewpoint car park. Return to Roybridge, and on the way down note the sand deposits alongside the road – these formed on the bed of the lake.

Return to the A86 past Spean Bridge (in a deep, narrow, glacially eroded overflow channel), then take the A82 south to Fort William. A stop at Darwin's Rest roadside tearoom at Roybridge is recommended – this is run by the Lochaber Geopark, and has local geology maps and books. The geopark office in Fort William also has materials for sale, plus an excellent display that explains the features of the geopark.

Chapter 8

Fort William, Glenfinnan, Lochailort, Ardnamurchan, Strontian

Introduction

From Fort William, take the A830 'Road to the Isles' going west (Fig. 8.1). Stop at Corpach for a magnificent view of Ben Nevis. Other good views are from the Banavie war memorial or from Neptune's staircase, the series of locks on the Caledonian Canal, at [NN 1137 7693]. Along the side of Loch Eil, the road goes through the flat belt of the Moine outcrop, in the Loch Eil Group, as far as Glenfinnan, where the West Highland granite gneiss appears, and the rocks are then in the steep belt, and the Glenfinnan Group. In the flat belt, sedimentary structures are preserved in low strain zones.

Locality 8.1 Fassfearn, Loch Eil

If time allows, take the small side road on the right, half way along Loch Eil, to Fassfern, and stop briefly at the bridge [NN 0209 7899]. Ripple marks on horizontal bedding planes of quarzite can be seen from the bridge. Continue in the same direction over the bridge and rejoin the A830, turning right for Glenfinnan, Lochailort and Mallaig. West of Glenfinnan the rocks are strongly folded, vertical and cut by numerous veins of granite and pegmatite, creating a very rugged landscape.

Locality 8.2 Glenfinnan

West of Glenfinnan, park at the lay-by [NM 8596 8152]. Here there are cuttings showing granite pegmatites with very large muscovite flakes. Climb up above the lay-by to the bare, ice-scoured rock pavement, where there are extensive clean exposures of coarse grey quartz–feldspar–muscovite schist, strongly folded and cut by many thick, white quartz–muscovite pegmatites containing large xenoliths of schist (Fig. 8.2).

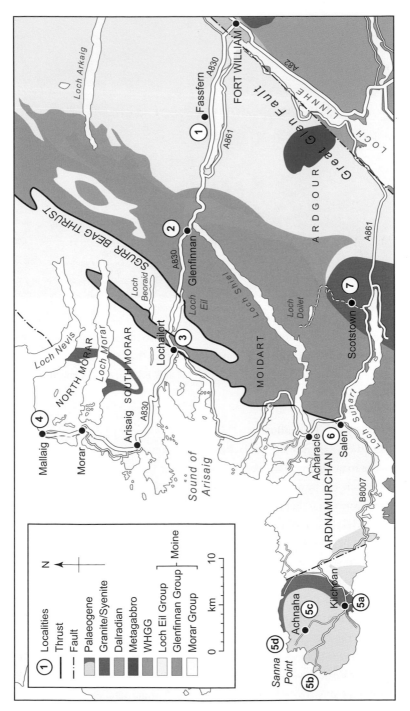

Figure 8.1 Route map with geology: Fort William to Ardnamurchan.

Figure 8.2 Glenfinnan: quartz–feldspar pegmatite cutting folded Moine gneiss.

Continue west along the narrow Loch Eilt road, and about 3km west of Lochailort, park on the wide grassy shoulder at the top of the hill, opposite some clean vertical rock cuttings [NM 7453 8326]. Cross the road with care.

Locality 8.3 Lochailort

Here there are a number of black dolerite dykes and horizontal sheets (sills) in the road cuttings. The vertical grey Morar Group Moine quartzites and schists are cut by quartz veins. Note how the dykes have a fabric and are cut by narrow zones that offset the horizontal sheets (Fig. 8.3). In places, some of the dykes cut cross bedding in the quartzites (Fig. 8.4).

At Lochailort, a detour to Mallaig can be made (18km). Beyond Arisaig there are stunning views west to the Sgùrr of Eigg and the volcanic mountains of Rum. Alternatively, go southwest on the A861 towards Salen.

Locality 8.4 Mallaig

Stop at the roadside car park just south of Mallaig [NM 6731 9681] and examine the Moine rocks (Morar Group) on the shore. The rocks are steep to vertical, very coarse silvery-grey schist and flaggy grey quartzite, with large muscovite flakes and abundant red garnets. A strong lineation

Figure 8.3 Lochailort: dykes cutting across Moine quartzite.

Figure 8.4 Detail of Fig. 8.3, showing cross-bedding in quartzite, to left of igneous sheet.

is obvious, and there are many thick quartz veins, as well as black dolerite dykes from the Skye swarm (55 million years old), frequently eroded into slots. From this point there are good views to Eigg, Rum and Skye (with the black Cuillin Hills of gabbro and the lower, rounded Red Hills of granite).

Drive to the harbour and park on the waterfront. Examine the tall rock cutting across the road. Here there are vertical dark grey pelites, slaty in places, and cut by many quartz veins. Some of the veins are pinched in the form of boudinage structure (Fig. 8.5). A steep lineation is visible at the east end of the section, as well as small garnet crystals.

Figure 8.5 Mallaig harbour: vertical Moine schist, cut by boudinaged quartz vein.

Drive past the harbour to the end of the road, 2km away at Mallaigvaig (Mallaig Bheag). Park at the last house, Arainn Mhòr, near the large turning place (avoid obstructing). Go through the gate and take the footpath to Mallaig Mhòr, on the way noting the vertical grey psammite (Morar Group). Sedimentary structures such as graded bedding and cross bedding are well preserved. Beds have the top to the west, which is the direction of younging. The slabs show a steep lineation plunging to the north. Across the bay are the mountains of Knoydart. Skirt round the bay at Mallaigmore (Mallaig Mhòr) and go down to the rocks on the east side of the bay, past the house [NM 6998 9770].

This locality is a Lewisian inlier exposed in the core of a large fold, the Morar antiform. The gneisses are very coarse, rich in biotite, intensely folded, with hornblende pods, rusty weathering patches and pink pegmatite segregations. It is not particularly easy to distinguish the adjacent Moine rocks. They are light grey, very coarse schists with quartz segregations, and are roughly banded with a gnarled or knotted appearance, due to numerous small-scale folds. Deformation has obscured the junction, but it is probably a sheared unconformity.

Return to Mallaig (accommodation, shops, tourist information, toilets, restaurants, railway station, ferries to Skye and the Small Isles) then drive back to Lochailort.

Locality 8.5 Ardnamurchan

From Lochailort, take the A861 road towards Kinlochmoidart and Acharacle. As far as Salen, this road goes through Moine schist. At Salen, take the B8007 road signposted for Kilchoan. On the way, note the brown-weathering Palaeogene dolerite dykes cutting the schists. From Glenborrodale, the Palaeogene lavas on top of Beinn Bhuidhe on the north side of Glenmore Bay can be clearly seen, with the characteristic horizontal step-like features of individual lavas. The lavas themselves are encountered some 3km to the west, with a notable change in scenery from bare rocky hills of schist to grassy slopes on the lavas. Beyond this point, the road becomes very narrow and winding, with few passing places, so progress will be slow. From the viewpoint above Camas nan Geall [NM 563 616] there are excellent views of Ben Hiant (528m, Fig. 8.6), the earliest of several vents belonging to the Ardnamurchan volcanic complex. Ben Hiant is made of dolerite, and it dominates the landscape of the whole peninsula. The headland of McLean's Nose is made of a volcanic breccia that filled the Ben Hiant vent. Continue west to the village of Kilchoan [NM 490 638] for accommodation and services. There is also a hotel at Sonachan [NM 453 664] some 3km northwest of Kilchoan on the B8007 road.

Ardnamurchan represents one of Britain's most spectacular classic sites of geology, as it contains the eroded cores of three overlapping volcanic centres. The concentric ring structure of Centre 3 is clearly represented in the landscape, as well as being very accessible. The volcanoes were intruded as ring structures into earlier basalt lavas, and they are practically contemporaneous with similar structures on the neighbouring Isle of Mull.

Figure 8.6 Ben Hiant: volcanic edifice guarding the entrance to the Ardnamurchan peninsula.

During the Palaeogene some 60 million years ago, a large plume of magma appeared in the north Atlantic Ocean, under what was later to become Iceland, related to ocean floor spreading and the rifting apart of Europe from North America. A north–south arm of this rift formed under what is now the Inner Hebrides region, and, where pre-existing faults intersect this rift, central volcanic complexes emerged, puncturing through the rigid basement of Lewisian gneiss, Torridonian sandstone and Moine schist. Thus, we now see St Kilda, Skye, Rum, Ardnamurchan, Mull, Arran and Ailsa Craig as the uplifted and deeply eroded cores of these ancient volcanoes. In the case of Ardnamurchan, the Moine Thrust is the crustal weakness that was exploited. The other centres were related to the Great Glen Fault, the Highland Boundary Fault and the Southern Upland Fault. Only a few of the more spectacular sites are included here. To appreciate the geology of Ardnamurchan requires a full week of fieldwork. See Emeleus and Bell (2005) for a detailed account of the geology, and the excursion guide by Gribble (1976, availabe on-line at Earthwise). There is also a very useful article on Centre 3 by O'Driscoll (2007).

Locality 8.5a Kilchoan

Follow the B8007 towards Kilchoan and turn left along the road signposted to Mingary Castle car park [NM 5005 6360], then walk down the track to

Figure 8.7 Kilchoan, Ardnamurchan: cone sheet (dipping to right) cutting across Moine schist (dipping shallowly to left).

the shore. The castle [NM 5030 6315] is perched on two horizontal sills that intrude Jurassic shales, causing them to be baked hard. The upper sill is a pale greenish-grey craignurite (silica-rich, a type of granophyre), while the lower one is a more basic dark brown dolerite. Both show vertical cooling joints, while the country rock shales are horizontal. To the east of the castle there are Jurassic limestones, weathering blue, with numerous white shells of the fossil lamellibranch *Gryphaea*, or 'devil's toenails', then there follows red Triassic sandstone above Moine schist. At Mingary pier on the B8007 road [NM 4941 6268], dolerite cone sheets, dipping west and belonging to Centre 2, cut Moine schist (Morar Group) and Jurassic shales (Fig. 8.7). The cone sheets contain xenoliths (blocks) of Moine rocks that were broken off the wall rocks whilst the cone sheets were being intruded.

Locality 8.5b Ardnamurchan Lighthouse
Drive west from Kilchoan to Achosnich and then left to the headland and lighthouse (toilets and refreshments). The lighthouse, which stands on gabbro outcrops from the Centre 2 volcano, has been constructed from the Ross of Mull granite (1849, Alan Stevenson). Return east on the B8007 then go left (north) onto the Sanna road.

Locality 8.5c Achnaha (Ach na h-Ath)

Park in the lay-by at [NM 4729 6750] then walk north of the footpath for 400m to [NM 4713 6825] across tonalite, an intermediate rock. This stunning location is at the centre of a perfect circle of black hills making up the Great Eucrite (a type of gabbro, with pyroxene, olivine and feldspar), and forms Centre 3, the youngest of the eruption centres (Fig. 8.8). The actual location itself is a small intrusion of monzonite. It is quite awe-inspiring to stand in the centre of a 55 million year old volcano.

Continue northwest for 2km to Sanna Bay, and park at [NM 4479 6938].

Figure 8.8 Ardnamurchan: part of the circular outcrop of the Great Eucrite intrusion; Rum, Eigg and Skye in background.

Locality 8.5d Sanna

From Sanna beach [NM 4576 7006] there are views to the Great Eucrite wall. The location is on the Centre 2 layered gabbro, a coarse khaki-weathering rock with a pitted surface due to erosion of soft olivine crystals. Layering is well seen at [NM 4407 7001]. Notice the smooth whalebacks here, caused by glacial erosion. In the small sandy cove at [NM 4599 7025] is the boundary between Jurassic sediments and the gabbro, with several basic dykes cutting the shales, disrupted by the intrusion. Across the bay are the Glendrian caves, formed by the weathering out of dykes that cut volcanic breccia. Out to sea there are excellent views of Rum, Eigg, Muck, Skye, Coll (Lewisian gneiss) and the Outer Hebrides. East of here the rocks are Upper Morar Psammites (Moine).

A possibility would be to take the Kilchoan ferry to Tobermory, then drive to Craignure, cross to Oban (7 daily, 45 minutes), and start Chapter 10. Otherwise, return to Salen and stop at the junction with the A861 in the car park by Sunart crafts (café and toilets) [NM 6905 6461].

Locality 8.6 Salen

Roadside cuttings here, beside the signpost for the single track road, are in vertical rusty weathering Glenfinnan Group Moine schists. The quartz–biotite–garnet schists are very coarse, with euhedral (perfectly formed) garnets 2cm across. About 4km east, stop on the lochside just by a small stream, at [NM 7329 6300]. Moine rocks here are cut by an east–west quartz dolerite dyke 10m wide, and showing two sets of perpendicular cooling joints, as well as vesicles (gas bubble cavities). This dyke is probably early Permian in age, i.e. older than the Palaeogene dykes of the Mull swarm.

Follow the A861 east along Loch Sunart and stop 3km west of Strontian at [NM 7879 6114].

Locality 8.7 Strontian

Exposures on the shore here are in the Strontian granite, 425 million years old. It is a composite intrusion, formed of an outermost (oldest) tonalite, then a more silica-rich granodiorite, followed by an even more silica-rich biotite granite. It is oval in shape, with a north-northeast trend, and was intruded along a fault that splayed off the Great Glen Fault, which was exploited by the intrusion (Hutton, 1988; Stephenson *et al.*, 2000). The granite has a weak foliation, resulting from deformation towards the end of crystallization (Fig. 8.9). Parallel to the foliation are elongate trails of xenoliths of Moine country rocks and dark hornblende-rich appinite, another igneous rock that is found nearby. It cuts across the Loch Quoich line (which is parallel to the Moine Thrust), with steep Moine rocks to the west and flat to the east, and this structure may have played a part in controlling the location of the intrusion. Previously it was thought that the Strontian granite was paired with the Foyers granite to the northeast along the Great Glen Fault, suggesting that the fault had moved to the left

Figure 8.9 Strontian granite with flattened elliptical diorite xenoliths.

by 80km. However, there are too many differences in composition, age and method of intrusion to sustain this notion nowadays; also, the total displacement on the fault is much greater than the 80km that separates these two granites.

From Strontian village, take the minor road north past Scotstown and Ariundle (famous for its ancient oak woods, a National Nature Reserve; the oak was used to make charcoal for the Bonawe iron furnaces in the eighteenth century). Near Bellsgrove Lodge the exposures are of tonalite (with plagioclase feldspar, biotite and hornblende). At the top of the hill [NM 8467 6660] there are folded banded basic lenses and boudinaged dykes in a granite gneiss, which is affected by highly irregular wispy folds lying in several directions, as well as patches of pegmatite (Fig. 8.10). The rocks here belong to the Ardgour gneiss, part of the 870 million year old West Highland granite gneiss, formed by partial melting of Moine rocks. This is an important date, in that it puts a limit on the age of the deposition of the sediments that make up the Moine succession, since the granite cuts Moine rocks. It was deformed and metamorphosed to a gneiss during Caledonian events.

Figure 8.10 West Highland granite gneiss.

Disused baryte workings are abundant downhill from the col, near the road at [NM 8380 6591]. Mineralization here at the contact between the Strontian granite and the gneiss includes old lead–zinc mines (18–19th centuries) in east–west veins, and the spoil heaps contain shiny grey galena cubes (lead sulphide), pale pink baryte veins, large white calcite crystals, and brecciated granite (Figs 8.11, 8.12). Baryte was extracted in the 1980s for use as a drilling mud in the North Sea. More recently, the Strontian granite has been exploited in the superquarry opposite Oban and Lismore, much of it used in the construction of the Channel Tunnel. The granite is also famous as being the type locality for the mineral strontianite (strontium carbonate, $SrCO_3$, discovered in 1791), from which the element strontium was isolated by Sir Humphry Davy (1778–1829) in 1808.

Return to Strontian and drive east, passing the granodiorite portion of the intrusion, as far as the A884 junction, then, if there is time, take this road west, then south across the Morvern peninsula to Lochaline. The road follows the edge of the Strontian granite as far as Claggan, where it turns west through Gleann Geal (Gaelic: white glen) and crosses into vertical Moine rocks. Just north of Lochaline are outcrops of red Triassic sandstone

Figure 8.11 Strontian lead–zinc mines.

Figure 8.12 Strontian mines: excavation of main lead–zinc vein.

and conglomerate, followed by Palaeogene basalt lavas. At Lochaline itself, near the Mull ferry terminal, is the famous Cretaceous glass sand mine beneath lava cliffs. The sandstone is an exceptionally pure white quartz sand, with perfectly spherical grains, indicating that it was blown from sand dunes into a shallow tropical sea.

It is possible to go from Lochaline to Mull (Fishnish) and drive to Craignure, then by ferry to Oban, and continue with Chapter 10. Otherwise, return to Strontian and drive east on the A861 through the U-shaped valley of Glen Tarbert. Halfway along on the south side is a superb example of a corrie, and the floor of the valley is covered in hummocky moraine deposited by a melting glacier. The road then turns northeast and follows the Great Glen Fault along the shores of Loch Linnhe as far as the Corran Ferry. From the ferry there is a splendid view along the Great Glen. Cross the Corran Ferry (frequent crossings, 10 minute journey, no bookings). Rocks on the east side by the jetty are crenulated (crumpled) blue-grey schists belonging to the Eilde Flags, part of the Dalradian succession (base of the Lochaber Group). They are cut by several thin dykes. Drive northeast on the A82 road to Fort William (20km) for accommodation and the start of Chapter 9.

Chapter 9

Fort William to Easdale, Kilmartin, Tayvallich and Kilmory

Introduction

This chapter explores the Dalradian rocks of Argyllshire, including the famous Easdale Slates and the pillow lavas at Tayvallich. The geological structure of this part of the Highlands consists mainly of large northeast–southwest striking upright folds, two of the major ones here being the Loch Awe syncline and the Ardrishaig anticline. This has resulted in the pronounced 'grain' of the country, with steep ridges and narrow valleys between. Metamorphism in southwest Argyll is generally low-grade, and the original sedimentary and igneous nature of the rocks is well preserved. However, the grade is higher to the north, in the vicinity of Loch Leven. After the Dalradian rocks were folded and metamorphosed, they were intruded by igneous rocks, such as the Ballachulish complex.

The Ballachulish granite forms part of the 425 million-year-old Ballachulish igneous complex, intruded into the Appin and Argyll groups of the Dalradian, which had already been folded and metamorphosed prior to the intrusion. Shortly before the main intrusion, there was a phase of igneous activity, when appinite explosion breccias formed. Kentallenite is a member of this family of igneous rocks. Like many of the Argyllshire igneous complexes, the one at Ballachulish is zoned, with an earlier more basic diorite, followed by the main granite. A metamorphic aureole surrounds the complex, whereby the country rocks were converted to hornfels by heat from the intrusion. The characteristic mineral cordierite grew during the earlier, diorite phase, when the temperature reached an estimated 1100°C, much hotter than the granite intrusion temperature of 850°C. Judging by the large number of xenoliths and larger blocks of country rocks, it seems likely that the present erosion level is close to the roof-zone of the intrusion. Further details can be found in the excellent

Ballachulish field guide by Pattison and Harte (2001, available online at BGS Earthwise).

Locality 9.1 Ballachulish granite

From Fort William, take the A82 southwest, following the Great Glen Fault, past Onich. (Note that there is an excellent and easily accessible section through the Appin syncline along the shore at Onich, well described by Roberts and Treagus, 1977 and Treagus, 2009). Turn left then cross the bridge to South Ballachulish, go west under the bridge (signposted to Oban) and park near the hotel at [NN 0499 5963]. Note the views to the Pap of Glencoe, made of Dalradian quartzite, and westwards towards the Sunart hills.

Blocks of fresh granite with xenoliths make the walls opposite the hotel. Exposures under the bridge and at the top of the steps leading to the monument are in coarse, pale grey granite, showing vertical jointing, and with abundant small xenoliths of Dalradian schist (flat, dark grey, foliated) and diorite (round, black, coarsely crystalline). The diorite inclusions are related to the earliest phase of the intrusion. The Ballachulish igneous complex has been dated at 425 million years old.

Continue on this road for 5km to Kentallen, and park at the Hollytree Inn near the pier [NN 0140 5836] (Fig. 9.1).

Locality 9.2 Kentallen

From the car park, cross the road and go through the gate past the old railway station, then up the hill, following the cycle track to the left, to the viewpoint and Geopark information board for the kentallenite at [NN 0115 5787], with Ardsheal Hill on the left, and looking towards Loch Linnhe (and the Great Glen Fault) and the fine U-shaped glacial valley of Glen Tarbert. Since this is a SSSI, there can be no hammering of the rocks. Kentallenite is an attractive dark khaki-green, very coarse-grained igneous rock, with large black hornblende and biotite crystals (Fig. 9.2). It belongs to a group of intrusive rocks known as appinites (named after the Appin district), which were associated with an early explosive phase of igneous activity just prior to the intrusion of the Ballachulish complex. Return to the car park, then walk to the pier.

At Kentallen pier [NN 0135 5842] there are blocks of white feldspar–quartz–biotite Ballachulish granite. As at locality 1, the granite has many

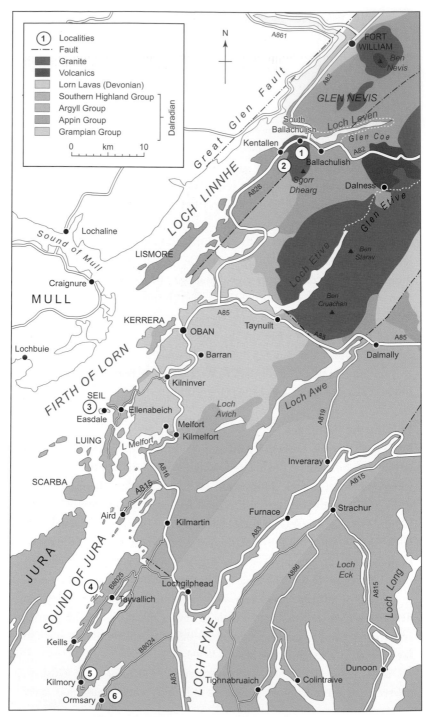

Figure 9.1 Route map with geology: Southwest Argyll.

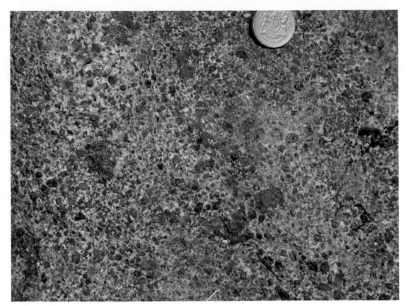

Figure 9.2 Kentallenite at Kentallen, the type locality.

xenoliths. Sheeted joints in the granite can be seen at Currachd Liath [NN 0243 5955], and in places there is patchy pegmatite. An offshoot of the granite cuts Appin Group metasedimentary rocks (slate, phyllite, quartzite, limestone), forming a contact-metamorphic rock known as spotted horn-fels, the spots being cordierite, an iron-rich aluminium silicate. This rock is smooth and extremely hard. To examine the contact between kentallenite and the Dalradian country rocks, walk past the hotel chalets along the old railway track, towards the headland. The first exposures at [NN 0115 5822] are in the metamorphic aureole or contact zone of the granite intrusion, and show prominent dark spots of cordierite on the very hard (baked) quartzite. On the shore at [NN 0107 5808] the even-grained kentallenite is in contact with the Dalradian rocks, and is very coarse, right up to the margin.

Continue south on the A828, stopping on the way at North Connel for a view of the Falls of Lora. This is a tidal race at the sharp edge of a glacially eroded rock basin. Cross the bridge and join the A85 to Oban. From Oban it is possible to take the ferry to Lismore to examine the Lismore Limestone (Dalradian), and from Gallanach to the Isle of Kerrera to see Devonian siltstones and lavas overlying a spectacular unconformity above folded Dalradian slates. Old Red Sandstone conglomerate with many

Figure 9.3 Dunollie Castle, Oban, constructed on ice-moulded basement of Old Red Sandstone conglomerate.

rounded igneous pebbles and boulders forms the shore at Ganavan, while slate is exposed on the shore in Oban town. On the way into Oban, there is a good view of Dunollie Castle, built directly onto a large rounded outcrop of Old Red Sandstone conglomerate (Fig. 9.3).

From Oban, take the A816 south over the Lorn lavas (Devonian in age), which show typical step features on the flat-topped hills. Thick individual lavas with vertical joints can be seen in places. At Kilninver go right onto the B844 and cross the bridge at Clachan onto Seil island (the 'bridge over the Atlantic'), then right at Balvicar and then to Ellenabeich, also referred to as Easdale (Fig. 9.4). Day trips to the Garvellach islands (stunning Precambrian glacial deposits, of world-class importance) can be arranged from Balvicar (Sealife Adventures www.sealife-adventures.com).

Locality 9.3 Easdale

The cliffs around the village are of Lorn lavas, Devonian in age (400 million years old; Fig. 9.4). Take the passenger ferry from Ellenabeich to Easdale Island [NH 7405 1710]. Quarrying for slate first began here in the twelfth century. At its peak, towards the end of the nineteenth century, annual production reached 10 million slates. The quarries finally closed in 1966. The

Figure 9.4 Easdale village: cliff of Devonian-age Lorn lavas, with slate outcrop in foreground.

main quarry on Easdale is 100m deep and was flooded by the sea in 1881. It has a 2m-wide brown-weathering vertical dolerite dyke marking one edge. Palaeogene dykes like this one (from the Mull swarm) commonly form margins to the quarries. The structure seen in this dyke is referred to as spheroidal weathering, where groundwater and rainwater have entered intersecting joints, making the surface progressively flake away like onion skins, leaving what resemble round cannon balls. The 5mm-thick slates can be seen in any of the numerous waste heaps and in the exposures. Note the fine crenulation (tiny folds causing a rumpling of the surface) and the abundant large cubes of pyrite or fool's gold (iron sulphide, FeS_2), some of which have weathered out to leave holes. At right angles to the cleavage surfaces it is possible to see another set of pyrite crystals, smaller and drawn out into elongate shapes. These are earlier than the perfect cubes, which are scattered randomly across the surface (Fig. 9.5). Light and dark colour banding is the result of cleavage planes intersecting the bedding. Where bedding and cleavage are parallel, larger slates were obtained. The Easdale Slate Formation belongs to the Argyll Group of the Dalradian, and lies above the Islay Quartzite. The slate islands of Seil, Easdale, Luing and Belnahua lie on the limb of a pair of large folds, the Islay anticline and the Loch Awe syncline, with the slaty cleavage running parallel to the axial planes of these folds. Information on these and other Scottish slates can be found in the report by Walsh (2000). If there is time, a visit to the adjacent island of Luing is worthwhile, with marked trails and a visitor centre explaining the

Figure 9.5 Easdale: slate showing earlier generation of pyrite in lineation feature, cut by later large pyrite cubes in random orientation.

geology. There are good views to the adjacent islands of Scarba, Jura, Islay, Colonsay, Mull, the Garvellachs and Insh, also the Glensanda superquarry in the Strontian granite.

Return to Kilninver and carry on south in the direction of Lochgilphead. There are important archaeological remains – stone circles, standing stones, burial chambers and many superb carved slate gravestone slabs – in the fertile limestone valley of Kilmartin Glen. The Kilmartin museum (with displays, restaurant, toilets and bookshop) is worth a visit. Gravestone slabs are exhibited in the churchyard, from where there is a good view along the glen to four river terraces, including the present-day floodplain; these represent different pulses of uplift since the retreat of the ice. Cup-and-ring marks, etched onto the surface of smooth, glaciated slabs of hard greenstone (metamorphosed lava) can be seen on the museum trail. A more extensive set can be found at Achnabreac (Figs 9.6, 9.7). A short visit to nearby Dunadd is also worthwhile (Historic Scotland). This is sign-posted just west of the main A816 road, and there is ample parking [NR 8369 9360]. Climb to the top of the small hill of greenstone (metamor-phosed lava) for a view across the 5000-year-old raised bog of Mòine Mhòr (Gaelic: the great moss), with ox-bow lakes in the abandoned meanders of the River Add (Fig. 9.8). It is a National Nature Reserve. The Iron Age fort of Dunadd is renowned as the site of the coronation of the kings of Dalriada

Figure 9.6 Achnabreac, cup-and-ring marks etched into greenstone (metamorphosed basic lava); note also the glacial striae, running left to right.

Figure 9.7 Kilmartin Glen; carved grave-stones in metamor-phosed lava.

Figure 9.8 View from Dunadd to Mòine Mhòr, with abandoned meanders in River Add.

(from which we have the geological term Dalradian), with a carved boar and an excavated footprint and water basin. An inscription in ogham script has been dated as eighth century. Achnabreac, mentioned above, is just off the A816, some 3km north of Lochgilphead, at [NR 8573 9065].

The last three localities in this chapter cover aspects of the geology of Knapdale, and involve long drives along three adjacent peninsulas and back again. Useful maps are BGS Sheet 28E, and OS Explorer 357 and 358. The overall structure of this region is the large Loch Awe syncline, complicated by the Tayvallich syncline, the Loch Sween anticline, the Kilmory Bay syncline and the Ardrishaig anticline (through Loch Caolisport) – folds of the same age but of lower amplitude and forming a sort of concertina of folds with parallel axial planes. All the rocks, which are part of the Argyll Group, are steep to vertical and strike northeast–southwest. Variations in resistance to erosion of the different rock types (limestone, greenstone, grit and slate), combined with the strong cleavage and overall fold structure, have given rise to the highly indented coastline, to the extent that some of the long fingers of land are practically islands. There are few roads cutting across strike, hence the long drives.

The localities are all GCR sites, described in meticulous detail by Tanner in Stephenson *et al.* (2013). Further information can also be found in the Dalradian guide to the Southwest Highlands by Roberts and

Treagus (1977, available online at BGS Earthwise), and Roberts (1998), while Anderton (1985) presents a comprehensive account of the connection between sedimentation and tectonic activity in the Dalradian. A very important feature here is the Tayvallich Volcanic Formation, which has provided a date of 600 million years for the igneous activity responsible for intrusions of sills, and underwater extrusions of lavas. This is one of very few precise dates for Dalradian rocks.

Locality 9.4 Tayvallich pillow lavas

At Cairnbaan, turn right onto the B841 (or left after visiting Achnabreac), following the Crinan Canal (at right angles to the regional rock strike), then take the B8025 along the narrow peninsula for 8km south of Tayvallich. The axis of the Tayvallich syncline runs the length of the peninsula. The green fertile areas are underlain by the Tayvallich Limestone, while most of the upstanding ridges are of greenstone (also referred to on old maps as epidiorite, a metamorphosed lava, the colour being due to the presence of the green minerals epidote and chlorite).

Drive through the village of Keillmore to the road end (through two unlocked gates – there is public access) and park in the lay-by at [NR 6895 8029]. The rocks on the beach just here are blue-grey limestones with rusty-coloured (iron-rich) dolostone patches and typical weathered-out hollows. Now walk past the house to the jetty at [NR 6886 8070], where sheared green porphyritic lava is exposed. Return to the farm track, then go through the gate and follow the clear broad track to the next gate, after which the track becomes less distinct. Follow it to [NR 6915 8104], then left towards the coast through a gate (facing Jura: the Paps are made of Dalradian Jura Quartzite), go through the gate diagonally opposite, in the corner of the field at [NR 6907 8125]. Walk along the top of the ridge to [NR 6927 8164], parallel to the wall, for 200m to its end at a narrow inlet, where it joins a second wall at right angles. Clean exposures on the rock platform at the water's edge present a wonderful three-dimensional view of a pile of 1 metre elliptical dark green to black pillows of coarse-grained spilite lava (a sodium-rich relative of basalt), at [NR 6938 8182]. These represent submarine lava flows. Note the concentric zoning, and the large white feldspar crystals internally, especially at the base of the pillows, whereas the edges are very fine-grained, i.e. the margins were chilled on contact with seawater (Fig. 9.9). The distinctive pink rock on the point at

Figure 9.9 Tayvallich: elliptical pillow lavas, showing concentric bands of gas bubbles

[NR 6955 8221] is a sill, and belongs to the same volcanic episode. It is referred to as keratophyre, a feldspar porphyry containing large sodium plagioclase feldspar phenocrysts. The tops of some of the pillow lavas have flattened fragments, which may have been shards of volcanic material swept off the surface and deposited on the pillows. Altogether, there are around 3000m of lavas, ash and sills in the succession, indicating a significant amount of volcanic activity at the time. This is a classic site, since the accurate age of 600 million years for the lavas gives a reliable date for this part of the Dalradian succession, at the top of the Argyll Group (Table 2.5). In plate-tectonic terms, the lavas represent the rifting of continental crust, an event that was to lead eventually to the rupture of the Rodinia continent and the formation of the Iapetus Ocean.

Return to the road, go past Keillmore House, through the first gate, and examine the roadside exposures of the vertical, sheared bluish-grey conglomerate – the Loch na Cille Boulder Bed, at [NR 6899 8031]. The bed can be followed along the peninsula of Rubha na Cille, and is well exposed beside the ruined electricity huts at [NR 6876 8019]. Here, it is vertical and contains clasts that are flattened into elliptical shapes, in a fine-grained pale green matrix. This bed has been the subject of controversy. It may be a submarine lava flow or a volcanic mudflow (termed a lahar, an Indonesian

Figure 9.10 Loch na Cille Boulder Bed, Keillmore – a possible Precambrian glacial deposit.

word). However, on the hillside just above at [NR 6909 8046] is an exposure with possible dropstones, which may have been deposited from the base of a melting iceberg. The boulder bed includes many fragments of greenstone, altered to the mineral epidote, whilst the matrix is full of quartz and small angular pebbles (Fig. 9.10). It has been dated at 580 million years old, the same age as the Macduff Boulder Bed on the Banffshire coast. At that time, Scotland was close to the South Pole and may have been in the grip of the late Precambrian Varanger glaciation (named after the Varanger Peninsula, Norway). Towards the point, the flat rock platform at [NR 6858 7975] has a clean exposure of large, elliptical bright green pillows showing concentric rings, as at the exposure to the north. The green colour is the result of alteration of the lava to epidote and chlorite by the action of water-rich fluids. This has also resulted in the rock being now very soft, and many of the pillows are eroded out (Fig. 9.11). The fine material surrounding the pillows is also rather pinkish and altered; this was originally volcanic ash and glass that formed at the same time as the lavas were erupted under water.

Keills Chapel [NR 690 806] contains one of the best collections of Celtic carvings in the West Highlands. These include the ninth century Keills Cross and grave slabs dating from the fifteenth century. Local slate and greenstone were used for these slabs.

If there is time on the way back to Tayvallich, another aspect of the lava succession can be seen on the coast at Port nan Clach Cruinn (although it is rather a long walk). Park off the road at Barrahormid, near an old quarry

Figure 9.11 Tayvallich: pillow lavas altered to epidote–chlorite rock by late-stage water-rich fluids.

[NR 7181 8395] and follow the farm track, which has a rock cutting at the first corner, showing greyish-green phyllite with a pronounced cleavage, and containing volcanic clasts. Continue to the ridge at North Ardbeg, near the cluster of ruined houses, climb the low fence, then descend via the gully to the narrow rock platform on the shore, beneath a low cliff. At [NR 7094 8461], between Port an Sgadain and Port nan Clach Cruinn there is a very interesting section of flattened green pillow lavas against finely bedded muddy brown-weathering limestone, showing typical solution hollows. A pronounced lineation can be seen in the limestone. The lava tops are full of gas bubbles, and are described as vesicular. At the junction between the limestone and lava, there are many thin quartz veins parallel to the bedding. Tongues of limestone appear to interleave with the lava, indicating that the lavas were erupted into soft sediment, pushing and disrupting the mud as they moved along the seabed, and forming pillow shapes. This locality was used by Peach (of Northwest Highlands fame) to illustrate that the rocks were right way up. He was also the first to report the presence of pillow lavas in Dalradian rocks.

Return to the car and drive north through Tayvallich and on to Barnluasgan and take the minor road at [NR 7900 9110] that leads down Loch Sween to Kilmory, past Castle Sween. Park on the wide verge near

Kilmory Chapel (with grave slabs made of epidiorite, or chlorite schist), at [NR 7032 7526] and follow the path that leads to the beach at Kilmory Bay.

Locality 9.5 Kilmory

The rocks in the bay belong to the Crinan Grit Formation, occupying the core of the Kilmory Bay syncline, with the underlying Ardrishaig Phyllites on the limb of this fold to the southeast, the junction being near Port Bàn. These two formations form part of the Argyll Group.

Head for Port Bàn on the south side of the bay [NR 7001 7403]. The rocks here consist of ribs or discontinuous beds of quartzite in fine-grained, strongly cleaved metamorphosed siltstone and mudstone. Cross bedding and graded bedding show that the rocks are right way up. The quartzites likely represent sand that was injected into mud and silt during deposition, possibly as a result of earthquake movement. Continue along the shore to the north, to [NR 7004 7435] where there are nice examples showing the relationship between bedding and the vertical cleavage parallel to the fold axial planes in the fine-grained bluish carbonate-rich mudstones (Fig. 9.12). Walk round to the north side of the bay to examine the Crinan Grits at [NR 6962 7472], opposite Eilean a'Chapuill. Here the rocks are

Figure 9.12 Kilmory: folds in Dalradian schist, showing fanning of cleavage around fold closure; coloured stripes are beds, steep features are cleavage planes (see Fig. 2.12).

pebbly conglomerates that grade up into coarse quartzite. These rocks were deposited in submarine channels, as fans of pebbles and sand carried down into deep water by turbidity currents. The sharp edge of one of these sandstone channels can be seen to the north, towards Port Liath, at [NR 6979 7540].

Return to the vehicle and drive back to Cairnbaan and Lochgilphead, then south on the A83 road, past Ardrishaig, and take the B8042 (north of Inverneill, at [NR 8517 8196]), southwest along Loch Caolisport in the direction of Kilberry.

Locality 9.6 Port Cill Malluig

Take the coast road and park at [NR 7230 7012] at the entrance to a track to the beach, just before the road bends inland; follow the track to the rocks on the shore. The fine-grained, greenish-grey to light-brown weathering rocks belong to the Ardrishaig phyllites, part of the Easdale Subgroup, and lie on the eastern limb of the large Loch Awe syncline. A narrow rock platform starts at [NR 7212 6999]. All the rocks are strongly folded, with a well-developed cleavage parallel to the axial planes. Excellent examples of folds and cleavages can be seen at [NR 7170 6956]. Fold amplitudes

Figure 9.13 Port Cill Malluig: folds in Dalradian rocks, showing different fold shapes and wavelengths in thick quartzite and thinner slate beds.

vary according to rock type and bed thickness, with the quartzites forming round, buckle-shaped folds that maintain the bed thickness. Finer-grained muddy rocks, on the other hand, are deformed into smaller-scale folds with limbs that are closer together (Fig. 9.13). This feature relates to different strength or competence of the beds during folding – quartzite is more competent and the beds maintain their thickness. On the other hand, the fine-grained muddy rocks are 'weaker' or less competent, and fold in such a way as to thicken up in the fold hinges. Bedding is easily identified by the colour change, and the cleavage is seen to fan around the fold closures (Fig. 9.12). Cleavage is much more strongly developed and intense in the fine-grained rocks compared to the quartzite beds, where it is more spaced out. From here it is possible to follow anticlinal fold hinges in the quartzite, plunging at 30 degrees to the northeast, as rounded 'whalebacks', swinging around slightly as the fold hinges are curved.

Chapter 10

Ben Nevis, Glen Nevis, Ballachulish, Glencoe to Loch Lomond

Introduction

The area to the south of the Great Glen Fault is dominated by large granite bodies (plutons) intruded into a basement of Dalradian schist and quartzite (Fig. 10.1). Related to the deep-seated intrusions are surface lavas and ash beds, fed by a huge number of dykes. Late Caledonian granites (also known as the Newer Granites, 425–400 million years old) are abundant in Argyllshire. They formed as a result of partial melting at the base of the crust, thickened due to north-directed subduction. Molten rock then rose upwards along structures that were inherent weaknesses and were exploited by the buoyant magma. Such structural controls included intersections between northeast–southwest faults of the Great Glen system and older deep shear zones or lineaments in the basement, trending northwest–southeast. For example, the Etive granite occurs where the Cruachan lineament is cut by the Etive–Laggan shear zone; the Glencoe volcano is located at the intersection between this same shear zone and the Glencoe line, while the Rannoch Moor pluton is controlled by the Ericht–Laidon fault and the Rannoch Moor line. Nearly all the plutons have a history of intrusion that starts with more basic diorite followed by a sequence of progressively more silica-rich (granitic) younger pulses of magma. Vast dyke swarms accompanied the plutons, and these fed surface lavas, most of which have been eroded away, except for some at Ben Nevis and in the Lorn district near Oban. Volcanic activity would have been violently explosive, judging by the composition of the lavas, ash and welded tuffs (ignimbrites, resulting from burning clouds of ash). Details of the research can be found in the map and guide by Kokelaar and Moore (2006).

From Fort William, drive northeast on the A82 for 5km to the Torlundy car park for the gondola station [NN 1720 7740]. Leave enough time for a

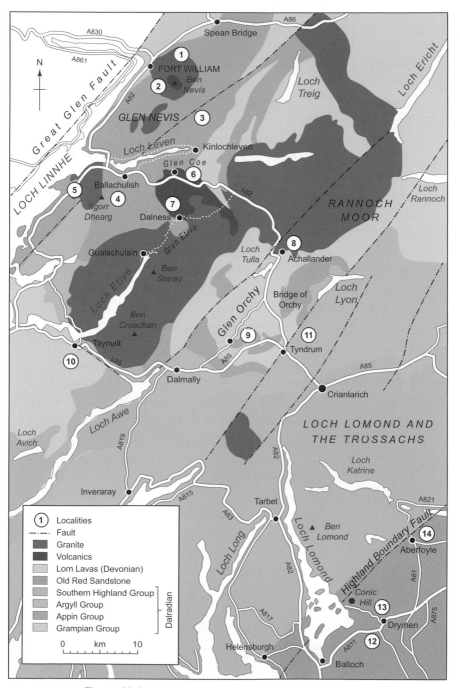

Figure 10.1 Route map with geology: Glencoe to Loch Lomond.

two-hour walk at the top of the lift. The main purpose is for the panoramic views, so clear weather is needed for this trip.

Locality 10.1   Fort William Gondola – Ben Nevis granite

On arrival at the upper station, follow the north path for 500m to the viewpoint at Sgùrr Finnisg-aig [NN 1890 7622], at 664m. From here there are views looking west to Loch Eil and east to the Grey Corries with their white scree of Glencoe Quartzite. The overall landscape to the west and northwest is of a dissected plateau, i.e. all the mountain tops are at the same height, with glacially excavated valleys between. Coarse, even-grained white-weathering granite is found at the top, with xenoliths of finely laminated grey Dalradian schist and folded blue limestone, cut by granite veins. Return to the station and take the other path that leads to the viewpoint at Meall Beag [NN 1784 7524], 1km to the west. Exposures here are in the granodiorite of the Ben Nevis intrusion. The andesite cliffs of the Ben Nevis ridge are visible from here, with the boulder field below.

Return to Fort William and go left at the roundabout along the minor road (signposted) to Glen Nevis and park at the visitor centre, [NN 1228 7302].

Locality 10.2   Ben Nevis

This excursion follows the so-called tourist route to the summit of Ben Nevis (1344m), but it should be borne in mind that this is a strenuous walk and requires stamina as well as the usual safety precautions. In order to appreciate the spectacular views, it is recommended to undertake this excursion only in good weather, and early in the day. On average, the summit is clear of cloud on only 10 days a year. Bearing in mind that the weather can deteriorate suddenly, and that snow lies at the top for long periods, it is vital to go equipped with warm and waterproof clothing, and some food and drink. Stout footwear is essential, as the upper section of the path is quite rough. Use the Explorer sheet 392 OS map for this walk, as it shows the footpath very clearly. Figure 10.2 is a view of the ben from Corpach.

Ben Nevis is a volcanic ring complex, 400 million years old, consisting of four consecutive intrusions that cut Dalradian schists, and capped by 700m of andesite lavas. (Note that Williams (2016) describes the lava sequence as trachyte and dacite, which are similar in composition to andesite, but with

Figure 10.2 Ben Nevis from Corpach.

rather more silica.) The first intrusion was diorite, followed by granodiorite, then two granites; i.e. the composition became more silica-rich with time. The first of the true granites is porphyritic, with large pink potash feldspar crystals in a finer matrix, and is cut by a swarm of dykes. This outer granite and the dykes are then cut by a later even-grained granite. Dykes are absent in the second granite. The lavas are Devonian in age and were erupted at the surface from a magma chamber immediately beneath. They form a plug that sank 600m into the younger granite as it was cooling and crystallizing, by a process known as cauldron collapse or subsidence. During the intrusion of the igneous rocks, the surrounding Dalradian country rocks were affected by heat and hardened to form hornfels, with new minerals growing in place of some of the original material in the metasedimentary rocks. The classic interpretation of the Ben Nevis igneous complex as a collapsed caldera formed by sinking of a central plug of andesite lavas down a ring fault has now been challenged, based on detailed remapping and 3D modelling (Muir & Vaughan, 2017; this paper is very well illustrated with maps, sections, photographs and models, in colour). These authors argue that the volcanic rocks forming the oldest part of the complex are not in fact a pile of andesite lavas downfaulted during the collapse of the caldera, but are actually a sequence of sedimentary and volcaniclastic material over 600m thick. Most of the material consists of blocks of ash and lava that formed as debris flows, and were derived from a volcanic centre that lay many tens of kilometres to the northwest of Ben Nevis. Additionally, there is no ring fault, and the material at the summit is now regarded as forming a large roof pendant of country rocks that once formed the local surface, and protected from erosion by being surrounded by the igneous rocks. In other words, the summit of Ben Nevis is close to the top of the igneous complex.

Start at the visitor centre car park [NN 1228 7302] and follow the signposted track past Achintee House and alongside the River Nevis. All the references are to exposures on the path itself – it is strongly recommended to adhere to the path and not to take short-cuts across the zigzags. Dalradian limestone and grey schist are crossed on the lower slopes, but are poorly exposed. Granite on the path is first seen at [NN 1314 7223]. Circular hollows 10–20cm across are weathered-out xenoliths of schist. The red granite is coarsely porphyritic (pink orthoclase feldspar), but weathers to grey. It is cut by frequent thin, bright red to orange felsite dykes, a fine-grained rock with the same chemical composition as granite (Fig. 10.3). Some of the dykes are porphyritic, with black chilled margins. An excellent example can be seen at [NN 1355 7191]. Some xenoliths within the granite are cut by thin 1cm granite veins. Perpendicular joints cut the granite into cubes. There are also metre-wide khaki-brown dolerite dykes with many closely spaced joints, cutting the granite. Slabs on the made-up path are mostly porphyritic pink granite. A few are of brown schist and dark

Figure 10.3 Felsite dyke on Ben Nevis footpath.

blue, finely laminated limestone and hornfels (a hard splintery rock formed when schist is intruded by hot granite). At the wooden bridge [NN 1381 7184], 350m, the U-shaped valley of Glen Nevis can be seen, with several truncated spurs, and the quartzite forming the grey screes at the head of the glen. At the next bridge above this point at [NN 1394 7179], 374m, the porphyritic granite and a xenolith of dark schist are both cut by a narrow felsite vein.

There is a sudden change to much redder, non-porphyritic granite at [NN 1427 7193]. The path still has some blocks of porphyritic granite with a very fine matrix, and a smooth weathered surface. Also, there are blocks of volcanic breccia containing angular fragments – this weathers a very pale grey. At about 550m [NN 1425 7205], there is a view to the andesite exposures lying above the granite.

Andesite is found on the path above Loch Meall an t-Suidhe [NN 1470 7239]. Grey and pink scree cliffs of angular, shattered andesite fragments cover the hill above this point. From here there are views to Loch Linnhe (in the Great Glen Fault) and Ardnamurchan. If the weather deteriorates or cloud descends, return to the bottom from here, as the route becomes much steeper and the path is covered with rough angular scree.

Glacial erosion features are well seen at 1000m, from [NN 1531 7134] onwards, including corries, truncated spurs and hanging valleys with suspended waterfalls in Glen Nevis, and the arête (a ridge formed by two corries back-to-back) at the head of the glen. The Cuillin Mountains of Skye are also visible. At [NN 1561 7141], 1160m, the andesite breccia is very clear. Fort William can be seen below.

Andesite buttresses form the unsurpassed view from the summit. Note the smooth round top above the cliffs (indicating that the summit was covered by an ice sheet), and the numerous joints that give the cliffs their rugged shape. Andesite is the fine-grained surface equivalent of diorite, an intermediate igneous rock, and was erupted as lava during explosive volcanic activity. The Torridonian mountains of Applecross are visible, as is Skye.

Return to the bottom by the same path, paying at least as much attention on the descent, especially on the andesite scree, which can be quite unstable. This walk is also described by Williams (2016), with some excellent aerial photographs.

Now drive up Glen Nevis to the car park at Poll Dubh [NN 1448 6838].

Locality 10.3 Glen Nevis

At the bridge, the small waterfall at Polldubh has formed at the point where two vertical dykes have cut through the coarse pinkish-red granite, which here is in the form of vertical sheets caused by jointing. Erosion along the junction between the different rock types has created the lower falls. Beyond the bridge at [NN 1453 6847], the large number of boulders fallen from the crags are grey, finely laminated and crenulated schists, part of the Dalradian Leven Schists (Lochaber Group). A strong vertical cleavage is very obvious in the rocks on the hillside above.

Continue up the glen, along the very narrow road. Just beyond a sharp bend at Màm Beag [NN 1605 6864] are some roches moutonnées (ice-moulded rocks with a steep plucked face and a smooth back; Fig. 10.4). The rocks are vertical grey schists, weathering brown, with minor folds deforming the crenulations. On the higher slopes above are many more examples of whalebacks, with their steep walls facing down the valley. Just beyond here, the Nevis waterslide comes into view. Continue on and park at the end of the road just after the waterslide (Allt Coire Eoghainn) [NN 1676 6911].

In the vicinity are many large erratic boulders, transported by ice and dumped in the valley when the ice melted. Walk back to the stream and go up the hillside for about 50m above the footbridge. Most of the waterslide is in smooth pink granite slabs, weathering to a very pale colour, but deep

Figure 10.4 Glen Nevis: roche moutonnée, a glacially shaped feature on Leven Schist.

red when fresh and wet (keep away from the edge). Boulders of blue-grey hornfelsed limestone are scattered in the stream bed. The large outcrop of coarse vertical schist in the granite at this point [NN 1664 6916] could be part of the country rock, or a huge xenolith in the granite. It is coarse-grained and pale grey, and the junction has given rise to a waterfall and small pool below. Xenoliths of schist can be seen in the granite immediately above the contact. Return to the car park, then follow the path up the valley, past an enormous fallen block of schist with a pronounced crenulation cleavage, at [NN 1684 6912].

On the path by a bend at a small waterfall [NN 1714 6925] is the contact between country rock schists and the Ben Nevis granite. The granite is deep red, fine and even-grained, with closely spaced joints, while the silvery grey muscovite-rich schist has crenulations and a lineation formed by the crenulation fold axes (i.e. like corrugated cardboard). Cutting the vertical hard schists are small, thin vertical cream-coloured dykes 25cm wide, and narrow elliptical white quartz veins. The schists are in the heat-affected zone (thermal aureole) of the granite, and in places there are numerous small, pale pink garnets that grew in conditions of thermal metamorphism.

The valley suddenly narrows into a deep gorge at Eas an Tuill (Gaelic: the waterfall of the hollow) with huge boulders and many potholes in schist, then at [NN 1749 6885] it opens out into a wider floodplain as far as the waterfall of An Steall (Gaelic: the water spout). A strong rodding lineation is very evident here, and the schist contains large muscovite flakes, as well as iron-rich (slightly rusty) elliptical patches, possibly representing replaced cordierite crystals and small square black crystals of magnetite (magnetic iron oxide).

Quartzite appears on the river bank, from the viewpoint to An Steall [NN 1749 6885]. Cliffs around the upper glen are grey schist, then bedded quartzite at the 100m high waterfall, with schist above and to the right, the waterfall being at the boundary. Higher up, to the left of the waterfall, the white quartzite is folded into a large overfold. This locality is also an excellent example of a hanging valley, caused when moving ice scraped away the lower edge of a tributary stream and cut down more deeply into the main valley (also creating a U-shape in the lower part). Post-glacial uplift then contributed to this side valley being at a higher level. The stream is flowing out of a corrie (Fig. 10.5). On the way back down the valley there are good views looking to the waterslide flowing over massive granite slabs.

Figure 10.5 Glen Nevis: Steall falls tumbling over Glencoe Quartzite from a hanging valley above.

Figure 10.6 Ballachulish, Loch Leven, Pap of Glencoe (Glencoe Quartzite); slate quarry visible on right.

For additional excursions in this area, see the map and book by Williams (2016), also the walks described by Gannon (2012).

Return to Fort William and take the A82 to Ballachulish; cross the bridge, then go east for 4km to the visitor centre beside the Ballachulish slate quarry [NN 0850 5850] (with cafeteria, toilets and a bookshop selling the Geopark leaflets). Figure 10.6 shows the quarry in its setting near the entrance to Glencoe, with the Pap of Glencoe (made of quartzite) forming a prominent landmark.

Locality 10.4 Ballachulish slate quarry

The disused quarry has a tourist trail with information boards describing the history of quarrying. The first quarry in the area began operations in 1693. At the entrance is an excellent example of a dolerite sheet cutting across the slaty cleavage (Fig. 10.7A, B) Cleavage and bedding are not coincident here, as can be seen from the slightly darker and lighter bands of

Figure 10.7A Inclined dolerite igneous sheet cutting obliquely across cleavage of slates in Ballachulish quarry.

Figure 10.7B Ballachulish Slate quarries; note very steep cleavage, and excavation benches.

colour (= bedding) on the cleavage planes. Rusty weathering results from the numerous pyrite cubes (FeS_2, iron sulphide or fool's gold). Fresh slate has a blue colour, while the cleavage surfaces form shiny silver faces. White quartz veins are very common, and some of these are quite thick, as can be seen at the top of the quarry. The slates belong to the Ballachulish Subgroup of the Appin Group Dalradian.

At the end of the nineteenth century, 400 workers were employed in the quarry, which by then had been operating for over 100 years. Operations ceased in 1955. Maximum average production was about 15 million slates per year, with an absolute maximum of 26 million in 1845. Vast amounts (over 85%) of extracted material ended up as waste, dumped into Loch Leven immediately opposite the quarry. Most of the slates went to central Scotland for roofing. Significant resources remain, and there are occasional plans to reopen some of the quarries to extract matching material for repairs to historic buildings, but the costs would be extremely high. For more on Ballachulish, see the report by Walsh on Scottish slate quarries (2000); the BGS website also has an extensive photographic library showing slate types from all the Scottish quarries.

Drive west for 2km to St John's church [NN 0675 5845] and park outside the gates, then cross the road and go through the gate onto the shore path and follow it to the headland of Rubha Poll an t-Seillisdeire. Low tide is needed for this locality.

Locality 10.5 Ballachulish Slide

Ballachulish Slates crop out on the shore before the headland is reached. After crossing a basic dyke at the water's edge at [NN 0662 5878], the rocks seen on the shore are vertical beds of laminated yellow Ballachulish Limestone (lying beneath the slates), showing minor folds with 's' and 'z' shapes, which indicate that the limestone lies in a syncline (Fig. 10.8). On the headland itself [NN 0663 5882], near a second dyke, the limestone meets a quartz-rich unit of the Leven Schists, belonging to the older Lochaber Subgroup. The junction is very sharp and slightly folded. It has been interpreted as a tectonic slide – a type of low-angle normal fault that may have formed during sediment deposition that was then reactivated during folding, causing adjacent units to be moved apart, possibly by a large distance, and cutting out part of the rock sequence. Treagus (in Stephenson et al., 2013) describes this locality in detail. On the north shore of Loch

Figure 10.8 Ballachulish Slide at Loch Leven.

Leven, the same relationship between the Ballachulish Limestone and Leven Schist can be interpreted, but the actual junction (the slide) is not exposed (see the excursion by Pattison & Harte, 2001, where they explain the important historical significance of the Ballachulish Slide).

Continue on the A82 for 3km and turn right into the Glencoe visitor centre car park (National Trust for Scotland; restaurant, toilets, bookshop, exhibition) [NN 1130 5760]. Glencoe has some very complicated geology, as well as being challenging physically, and only a brief account of a few roadside stops can be presented here. To appreciate the geology more fully, refer to the outstanding map and guidebook by Kokelaar and Moore (2006), which offers excursions to some of the higher locations in Glencoe.

Locality 10.6 Glencoe

The road through the glen crosses a deeply eroded volcano that experienced violent eruptions during the Devonian Period, 420 million years ago, and eventually sank deep into the crust around a series of ring faults, by collapse of the central caldera. This cauldron subsidence, as it was called in 1907 by Clough, Maufe and Bailey from the Geological Survey, became famous worldwide, and Glencoe has been studied intensely since then, as a classic of its type. The Glencoe complex is cut by the Etive granite, which in turn is cut by a northeast–southwest dyke swarm dated at 415 million years. Overall, the volcano complex is elliptical, measuring 16km by 8km, the long axis being oriented northwest–southeast. Parallel to the short axis, many dykes have been intruded, which had the effect of exaggerating the elliptical shape. Two internal faults, also northwest–southeast, created a graben structure (a fault-bounded trough), which allowed material to sink downwards during repeated volcanic episodes interspersed with periods of erosion and sedimentation. When the surface structure of the volcano foundered and sank into the almost empty chamber below, a final pulse of magma was injected upwards into the ring-fault.

From the visitor centre outlook point there are excellent views east to the Aonach Dubh cliffs, guarding the entrance to the glen (Fig. 10.9). Note the distinct colour difference at the sharp junction between the

Figure 10.9 Glencoe, Aonach Dubh: cliffs of rhyolite (lighter) and andesite (darker) lavas and sills.

upper light rhyolite lavas lying on dark andesite and basalt sills. Rhyolite is the surface equivalent of granite, and was erupted in viscous (sticky, due to the high silica content) flows that usually preserve colour banding, sometimes folded during movement across the surface (Fig. 10.10A). The distinct gully running up An t-Sròn (the nose) is excavated in the ring-fault, and the hill itself is made of the ring intrusion granite.

Drive on for 3km and turn left onto the old road (no stopping or parking allowed), over a small bridge, to the Clachaig Inn, and park there for a view of Creag nan Gobhar [NN 1278 5676], with the deep vertical cleft known as Clachaig Gully or Eas an Daimh in Gaelic (waterfall of the stag), with a scree fan at the foot. This has been excavated along a fault. The rocky peak above (Sgòrr nam Fiannaidh), like the Pap of Glencoe, is Glencoe Quartzite, while the valley floor here is made of Ballachulish Limestone.

Return to the old bridge and onto the A82, then right for a few metres onto the rough road by Loch Achtriochtan and park near the bridge, then return on foot (take great care on the road through Glencoe at all times) to the small quarry at the bridge over the River Coe [NN 1375 5665]. Loch Achtriochtan lies on the ring fault, and this quarry is in rocks of the ring intrusion. On the hills above are low cliffs of a well-jointed, pale grey and yellow-weathering andesite lava, reddish in places, especially near the road bridge junction. It is extremely hard and brittle, and has a greenish-white surface when fresh. The obvious pink phenocrysts are feldspar and there are also smaller black augite crystals in a fine matrix.

Continue to the lay-by opposite Loch Achtriochtan and park [NN 1460 5713]. Walk uphill a short distance across the grassy slopes to the exposures of phyllites – silvery grey, flat slabs by a small waterfall at [NN 1458 5722]. These are the Dalradian country rocks (Leven Schist Formation) that form the basement to the Glencoe complex. Small black octahedra of magnetite define a lineation on the phyllite surfaces. Higher above this location, the vertical cliffs are andesite lavas of the Aonach Eagach ridge, dipping south into the centre of the caldera, and unconformable on top of the phyllites.

Continue east to the Three Sisters car park [NN 1710 5689]. From here there are views to the nearby black boulder slopes or debris cones, formed from material tumbling down the slopes of the Aonach Eagach ridge. As can be seen from the lack of vegetation, these are still actively moving, and they occasionally block the road. Opposite is Ossian's Cave, a hollow excavated from a dyke. The serrated nature of the ridges in Glencoe is due in part to

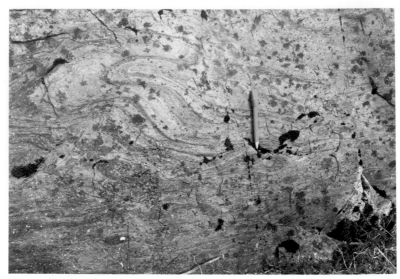

Figure 10.10A Flow banding in Queen's Cairn rhyolite lava, Glencoe.

the weathering out of vertical dolerite dykes, which are more easily eroded than the hard rhyolite and andesite. Looking west down the main valley, the edges of the smaller side valleys descending from the Three Sisters were sliced off by the glacier that moved down the glen; these are referred to as truncated spurs. Rhyolite lavas on Beinn Fhada are very obvious here. The Lost Valley (Allt Coire Gabhail) is a hanging valley eroded along a dyke and blocked by a rock fall from Geàrr Aonach.

Continue to the two lay-bys at a new bridge over a deep gorge beside a waterfall [NN 1836 5624]. Grey rhyolite is present here, strongly jointed and showing flow banding. Note the variable colour of the rock at the bridge – pink, grey, green, with white and pink phenocrysts (larger lath-shaped crystals of feldspar, lined up in the flow direction) and green clusters (Fig. 10.10A, B). The rock, the Upper Etive Rhyolite, has a very uneven grain size, and breaks into sharp angular shards. It is cut by a pinkish red micro-diorite dyke, seen in the stream bed at the base of the waterfall, beneath the bridge.

Stop at the last lay-by leaving Glencoe [NN 2129 5599], for a view back down the U-shaped glen, and to nearby Buachaille Etive Mòr (Figs 1.2, 10.11), which consists mainly of volcanic breccia followed by rhyolite lavas and ash, with a few thin shale beds containing plant stems of Devonian age. The precipitous cliffs towards the summit, Stob Dearg (1022m), result from

Figure 10.10B Basic xenolith in porphyritic rhyolite, Glencoe.

Figure 10.11 Ring of mountains surrounding entrance to Glencoe, with Buachaille Etive Mòr right of centre.

rhyolite being very resistant to erosion. Many of the deep gullies are eroded dykes. About a kilometre east of here, the outer fault line of the cauldron subsidence crosses from Altnafeadh and runs north of Glencoe through Stob Mhic Mhartuin, partly following the Devil's Staircase, part of the Old Military Road to Kinlochleven, on the West Highland Way. Between Stob Dearg and the ring-fault, the low ground is occupied by Dalradian basement, the Eilde Flags of the Grampian Group. The view east into the broad, flat Rannoch Moor is equally impressive (see Fig. 2.12). It is underlain by a granite that was deeply eroded during the Cenozoic (from 60 million years ago). Turn right onto the road past Buachaille Etive Mòr down Glen Etive and park at [NN 1993 5128].

Locality 10.7 Glen Etive

Granite slabs in the river bed with horizontal sheet joints can be seen at [NN 1993 5128]. This particular granite belongs to the Clach Leathad pluton. Several dykes crop out along the roadside – these are pale greenish-grey and porphyritic with white feldspar phenocrysts, and cut the granite. They belong to the Etive swarm of microdiorites.

Return to the main A82 road and turn right, past the ski road (from where there are good views to Buachaille Etive Mòr).

Locality 10.8 Rannoch Moor

Rannoch Moor was the centre of ice accumulation for the best part of two million years, during the Quaternary Period, and a number of glaciated valleys radiate outwards away from the moor. Ice flowing westwards down Glencoe was responsible for shaping the valley and truncating the spurs of tributary streams, some of which produced hanging valleys. The Rannoch Moor granite is cut by the fault intrusion of Glencoe, and so is older. Boulders of this granite are found within some of the volcanic conglomerates of Glencoe, indicating that the Rannoch Moor granite had been eroded and was exposed at the surface before the volcanic activity at Glencoe. Numerous large, round boulders of granite are scattered across the shallow lochs, moraines and peat bogs of the moor, the boulders being bleached to a pure white on the surface by acids in the peat, which convert feldspar to clay minerals. These boulders can be examined at the roadside by Loch Bà, and the typical boggy landscape is well seen at Lochan na h-Achlaise. To the west lie the granite hills of the Etive complex, the southernmost outcrop of which is Ben Cruachan.

A brief stop may be made in the large lay-by overlooking Loch Tulla [NN 3085 4518] for a view north and west to the granites, which contrast with the view south towards Dalradian rocks. A disused quarry opposite the lay-by shows granite with quartz veins cutting Dalradian quartzite (Grampian Group). Various sulphide minerals are present in the quartz veins. The valley floor around Loch Tulla is filled with glacial moraine hummocks.

About a kilometre south of Bridge of Orchy, a choice can be made for an optional detour to see the Falls of Orchy (Eas Urchaidh). Turn right onto the single-track B8074, opposite Ben Doran (Beinn Dòrain, made of Appin Group Dalradian, the low ground being of older Grampian Group rocks). Previously it was thought that the rocks here represented a tectonic boundary between Moine and Dalradian rocks, but it is now known to be where Grampian Group rocks give way to the overlying (younger) Appin Group in a normal sedimentary transition.

Locality 10.9 Glen Orchy (Gleann Urchaidh)

There is a lay-by in Glen Orchy opposite the iron bridge at [NN 2432 3206], 10km from the junction with the A82. Great care must be taken at this locality, as the rocks dip towards the river and are slippery – do not approach if the river is in spate. The rapids are formed of flat slabs of quartzite and garnet–mica schist, belonging to the Appin Group of the Dalradian. Garnet is mostly altered to green chlorite, but the original crystal shape is still evident. The main structure here is the recumbent (lying on its side) Beinn Udlaidh syncline (named after the hill to the east of the River Orchy), domed up by a later fold that caused a network of faults and fractures to form, including some that are mineralized (with lead, gold and silver). Downstream, at the dam [NN 2425 3186], folds in the quartzite plunge gently downstream to the south – this is the plunge direction of the main Beinn Udlaidh syncline. The rocks above the dam belong to the Leven Schists. On a small scale, folding has resulted in thinning of the quartz-rich layers to such an extent that their limbs have been stretched out to nothing, leaving just the cores. Larger-scale, flat-lying isoclinal folds are very abundant and clear along both sides of the river, which runs more or less along the axis of the syncline.

Continue to the gravel-covered lay-by just before the road leaves the river, at [NN 2467 3269], then go through the gate onto the rocks at the protective wall on the river edge. Here the quartzite has cross bedding, from which it is easy to deduce that the rocks are right way up. This area (a GCR site; see Tanner (in Stephenson *et al.*, 2013)) was one of the first places where sedimentary features were used to work out the overall structure of the rocks. Carry on along the gravel path to [NN 2472 3307], where there is an excellent section of flat-lying fold hinges like a pile of logs (mullions) in the clean white quartzite beds, just above the Eas a' Chathaidh waterfall (Fig. 10.12). A fuller excursion description can be found in Treagus (2009); but note that Treagus has the main folds as first generation, whereas Tanner proved that there are earlier structures, hence these are second folds.

Figure 10.12 Glen Orchy: pile of recumbent folds in Dalradian quartzite.

From here, follow the road southwest to meet the A85 at Inverlochy and turn right (west) to Dalmally and past the north shore of Loch Awe.

Locality 10.10 Cruachan, Loch Awe

A visit to the underground cavern of the Cruachan power station is worthwhile, if time permits. This is a pumped storage hydroelectric scheme, with water pumped to the dammed artificial loch in the corrie at the top of the hill (Fig. 10.13). During peak demand, the water is used to power the turbines 1km below ground, inside the granite (for information, see www.visitcruachan.co.uk). The Cruachan granite is part of the Etive complex and has an early diorite outer edge, followed by a much larger granodiorite. This in turn is succeeded by a porphyritic outer and an inner even-grained granite, at Ben Starav to the northeast. The Cruachan granite, dated at 400 million years old, cuts Glencoe volcanic rocks. Dalradian country rocks in the vicinity of the Pass of Brander (Ardrishaig Phyllite, at the top of the Easdale Formation) have been altered by heat from the granite for up to 2km away. The pass has been eroded along a fault.

The alternative route from Bridge of Orchy is to continue to Tyndrum (shops, toilets, restaurants, petrol). West of Tyndrum, there are many obvious excavations on the hillsides, marking the sites of former lead mines

Figure 10.13 Cruachan granite, with dam for underground pumped storage hydroelectric scheme.

and more-recent exploration for gold (1990, 1995, 2008). Mineralized quartz breccia veins containing galena (lead sulphide) and sphalerite (zinc sulphide) and other veins with gold and silver are associated with the Tyndrum fault, a 60km-long northeast–southwest fracture that belongs to the Great Glen system of faults, and has a left-directed lateral movement. Mineral veins (dated at 410 million years) cut through quartzites of the Grampian Group Dalradian rocks to the west of the fault.

Locality 10.11 Tyndrum mines

To visit the remains of the old lead and zinc mines, park at Tyndrum Lower railway station car park [NN 3276 3023], cross the railway (with care) and walk up the forest path past some ruined mine buildings and fenced-off shafts and adits to the spoil heaps at [NN 3192 3038], where it is possible to collect samples of galena and sphalerite in the quartz breccia host rock (Fig. 10.14). Note the view across the valley to the various trial workings on the hill opposite, and the glacial moraine hummocks on the valley floor, on both sides of the railway line. Treagus (2009) has a fuller description of this locality.

Several attempts at gold extraction have been made at nearby Cononish. Native gold itself is not seen, as the metal is present in pyrite (iron sulphide)

Figure 10.14 Tyndrum: mineralized breccia from abandoned lead and zinc mines.

and galena. Mineralization was due to circulating hot, water-rich fluids moving up through fractures in quartzite and schist, the heat source being the underlying granite. Economic studies show that there are at least half a million tonnes of ore with metal grades of 10 grammes per tonne of gold and over 40 grammes per tonne of silver. Other estimates suggest around 5.5 tonnes of gold and 28 tonnes of silver could be obtained. In 2016, the first gold was obtained from Cononish, extracted from previously mined material, and made into commemorative coins. A detailed study of the geological structure of this area and its ore deposits is presented by Tanner (2012), and useful information is available on the Scotgold Resources website.

Continue on the A82 down the west side of Loch Lomond, passing Dalradian slates. Many of the steep cuttings are covered in netting, and it is not possible to stop to examine the rocks except for some small exposures along the old road between Rubha Mòr and Rubha Dubh. These are the only accessible and safe places to see these rocks, but they are now quite overgrown. Bowes (in Lawson & Weedon, 1992) describes the localities, and Treagus (2009) offers a brief excursion to the Dalradian rocks on both sides of Loch Lomond, as well as the Highland Border Complex at Balmaha (locality 10.13).

Approaching Tarbet, Ben Lomond dominates the view on the east side of the loch. It is made of Ben Ledi Grit, a quartz-rich conglomerate in the Southern Highland Group, above the Aberfoyle Slate (Table 2.5). Scars on the hillside are the result of landslips in the 1990s. North of Luss, the loch is very narrow – it is a deep glacial valley in pale greenish-grey Dalradian phyllite of the Southern Highland Group, cut by numerous white quartz veins, and well exposed in steep cuttings on the roadside. This is in sharp contrast to the wide, shallow southern end of the loch, which is in Old Red Sandstone. The main structure here is the Tay Nappe, a large overfold that has put the slates upside down in the so-called Highland Border Downbend. Farther north, beyond Inverbeg, the Dalradian rocks are in the flat limb of this major fold. Originally, the rocks were mud, silt, shale and sand, meta-morphosed to phyllite, slate, schist and quartzite. Travelling south, the meta-morphic grade decreases from schist to phyllite then slate. This is reflected in the mineral content: silvery white mica in schist and green chlorite in phyllite. The axis between the steep belt and the flat belt occurs at Rubha Mòr, 2km north of Inverbeg. Within the flat belt, the rocks are upside down. Towards the south shores of Loch Lomond, the flat farmland is mostly underlain by Old Red Sandstone deposits. Looking east from Duck Bay, the chain of islands (Inchmurrin and Inchcailloch being the largest) in the loch, and Conic Hill above Balmaha (locality 10.13) on the east side mark the line of the Highland Boundary Fault. A dip in the skyline (Ben Bowie) to the southwest at the A82/B831 Helensburgh roundabout marks the location of the fault – the fault actually runs through the middle of the roundabout. Old Red Sandstone conglomerates on the south side of the fault have been pulled up into a vertical orientation during faulting, the south side having moved downwards. Along the fault itself is an igneous intrusion that has been altered to serpentinite by the action of fluids.

At Balloch, take the A811 (towards Drymen) and stop at Gartocharn.

Locality 10.12 Gartocharn

In Gartocharn, take the Duncryne road, then park on the left at [NS 432 856] beside the path signpost. Walk up the footpath through the wooded area with an old quarry on the left. The quarry is in an early Carboniferous volcanic plug of basalt and agglomerate, showing some columnar jointing. Continue to the top of Duncryne Hill (142m) [NS 436 859] for a view of Loch Lomond, towards Ben Lomond and Conic Hill (Fig. 10.15).

Figure 10.15 Loch Lomond and Ben Lomond (Dalradian) from Duncryne, showing Highland Boundary Fault marked by chain of islands, and continuing to Conic Hill on right.

Note how the loch narrows to the north, with the Caledonian mountains, while the southern part is much wider and flat, on the younger Old Red Sandstone. The islands crossing the south of the loch mark the line of the Highland Boundary Fault.

Continue to Drymen, then Balmaha on the B837 and park at the National Park visitor centre. Refreshments and toilets are available, and a geological leaflet of the area (also available to download from the Geological Society of Glasgow website).

The Highland Border controversy

Adjacent to the Highland Boundary Fault is a narrow zone of rocks belonging to the Highland Border Complex – from Arran through Bute, Balmaha, Aberfoyle and all the way to Stonehaven in the northeast. The rocks that make up the complex are weakly metamorphosed sediments (sandstone, shale and limestone), together with ocean-floor pillow lavas in a few places (e.g. Stonehaven) and fragments of ultrabasic igneous rocks that probably represent ocean floor crust and upper mantle – the Highland Border Ophiolite. Individual segments of the complex are small and strongly faulted, as well as being far apart from one another. The lack of identifiable fossils and sedimentary way-up evidence have contributed to various conflicting views on the origin and history of the complex: a controversy that lasted for over 100 years. Briefly stated, one view (that of the late Brian Bluck of Glasgow University) was that the complex represented an 'exotic terrane' that was tectonically placed between Dalradian rocks of the Highlands and the Old Red Sandstone of the Midland Valley. Detailed remapping of the entire area by Geoff Tanner (also of Glasgow University) demonstrated that the terrane model was unnecessarily complicated, and in fact the sedimentary rocks of the complex graded up from the top of the Southern Highland Group of the Dalradian. Tanner defined this section as the Trossachs Group of the Dalradian. Structurally above is the Highland Border Ophiolite, a slice of ocean crust that was thrust in along the edge of the Midland Valley terrane. This interpretation was confirmed in 2008 by an international team who visited all the relevant sites. At its simplest, the sequence proposed by Bluck is completely upside down relative to the one demonstrated by Tanner: the ophiolite is at the top and not at the base. The rocks above the ophiolite belong to the Garron Point Group, named after the locality at Stonehaven, and the Highland Border Complex now represents the Trossachs and Garron Point groups. For a full discussion of this topic, with maps, sections and photographs, and how the apparent paradox was resolved, see Tanner and Sutherland (2007). Henderson *et al.* (2009) describe and interpret the structure of the Highland Border Ophiolite, while Bluck (2015) deals with the Highland Boundary Fault and associated structures at Aberfoyle.

Locality 10.13 Balmaha: Highland Border Complex
From the visitor centre, walk along the lochside footpath towards Balmaha Pier and follow the signposted path uphill on the right through the trees, to Craigie Fort. Exposures at the top are of Lower Old Red Sandstone conglomerate and sandstone, dipping very steeply towards the southeast, having been tilted up against the Highland Boundary Fault. The well-rounded smooth pebbles and cobbles in the conglomerate are mostly of white and pinkish-grey Dalradian quartzite, and occasional granite. Many of the pebbles have fine cracks running across them, as a result of fault movements. As at Duncryne, there is a good view to Ben Lomond and the chain of islands in the loch. Return to the road and turn right to the pier at [NS 4155 9008], where the conglomerate can be examined again. Continue along the shoreline path and cross the metal footbridge, where

there is a clear section in almost vertical conglomerate beds. On the shore at [NS 4153 9098] the exposures are now in vertical, well-bedded red sandstone with thin conglomerate layers. The sandstone is then followed by another coarse rock at [NS 4144 9143], but note that here the clasts are of thin, angular, flat grey Dalradian schist, quite different from those in the rock at the pier. Since the clasts are angular, the rock is often referred to as a sedimentary breccia, rather than a conglomerate; i.e. the clasts had a shorter travel distance and so are not rounded. This exposure is of Upper Old Red Sandstone, which is unconformable on the Lower Old Red Sandstone – the Middle Old Red Sandstone is missing in central Scotland. On the shore at [NS 4095 9175], just before a small islet south of Arrochymore Point, the rock is a pale purple to pinkish-grey hard, gritty, coarse sandstone, marking the base of the Trossachs Group (top Dalradian). Although it is difficult to see, cross bedding can be found that shows the rock to be upside down, with a shallow dip to the north, and younging to the south. Folding has caused all the Dalradian rocks to be inverted here.

On the point itself, at [NS 4103 9194], is a pale grey, coarse crumpled rock, showing irregular foliation and shearing, cut by thin red veins of chert (or jasper, iron-stained silica) and silicified serpentine. In places the rock is very dark green, and is associated with a conglomerate. Red jasper boulders are quite abundant on the beach. Serpentine is well exposed on a low cliff-edge in the woods, just above the path [NS 4105 9196]. Here, the rock that forms the hill to the northeast above the visitor centre is a mixture of black and very bright green shiny serpentinite, derived from metamorphism of ocean-floor igneous material (Fig. 10.16A). It is part of the Highland Border Complex.

From here, walk to the Arrochymore car park, then right (take care) along the road towards Balmaha. On the right at [NS 4123 9200] is a high cutting, rather overgrown with moss, which can be peeled back to show the lustrous hard red chert rock of the Highland Border Complex (Fig. 10.16B).

Continue along the road until reaching a gate at [NS 5416 9133], then follow the path uphill through the woods. Red serpentinite conglomerate can be seen on the path, which weathers to give a very bright reddish-brown soil at the point where the path narrows and becomes steeper. At the top the path runs along the ridge of Druim na Buraich, with cliffs below, marking the edge of the Gualann Fault [NS 4178 9138]. The

Figure 10.16A Highland Border Complex at Balmaha: green serpentinite rock, representing altered igneous rock.

Figure 10.16B Balmaha: red jasper rock in Highland Border Complex.

yellow-weathering serpentinite conglomerate is well exposed on the path above the Old Manse, from where there are views of Loch Lomond, Inchailloch and Conic Hill (Fig. 10.17). From here, the path descends from the serpentinite conglomerate, then crosses onto Old Red Sandstone to the

Figure 10.17 Conic Hill, Loch Lomond: serpentinite altered to carbonate rock along Highland Boundary Fault.

Figure 10.18 Loch Lomond: view from Conic Hill to string of islands marking location of Highland Boundary Fault.

top of Conic Hill (361m). It is worth climbing this hill to view the line of islands marking the Highland Boundary Fault (Fig. 10.18). On the way back to the Balmaha visitor centre car park, there are many good exposures of red sandstone and quartzite conglomerate on the path, showing steep to vertical bedding.

Return to Drymen on the B837, then to Aberfoyle via the A811 and A81. At the end of Aberfoyle village, take the A821 uphill towards Dukes Pass, and park in the David Marshall Lodge visitor centre [NS 5202 0142]; parking is charged.

Locality 10.14 Aberfoyle: Highland Boundary Fault

The area around the lodge, as at Balmaha, lies within the Highland Boundary Fault zone, and the rocks to be examined belong to the Highland Border Complex, sandwiched between the Gualann Fault and Lower Old Red Sandstone conglomerate. Take the footpath behind the lodge, then go downhill and cross the stream by the bridge. The hard greenish rocks in the stream are vertical Dalradian grits containing angular quartz fragments, with a chlorite-rich cement. Continue on the track, to meet a path on the left, and make the steep climb on the former train track through the forest, up to Lime Craig quarry. High grassed-over spoil heaps mark the entrance to the quarry. On the left at the entrance is a small vertical dolerite dyke of Carboniferous age, intruded along the Gualann Fault. Above the dyke, also on the left, is a dark lustrous serpentinite, in various shades of purple, green and black, criss-crossed by many thin veins of yellow dolomite, pale green talc and dark green serpentine. Cross the fallen boulders with care to the back of the quarry to examine the conglomerate (Fig. 10.19). The boulders are elliptical, well-rounded, and arranged in an imbricate fashion (i.e. stacked up) along the vertical bedding. Many of the shiny red, hematite-stained quartzite boulders and cobbles have fractures cutting across and often displacing the various pieces of the boulders. The fractures, which relate to fault movements, have fused together and resealed the boulders. Cross over to the other side of the quarry to view what remains of the limestone. It is a very coarse, massive, orange-coloured rock that was extracted and taken down the incline railway to lime kilns at Dounans, at the bottom of the hill, to be burnt and used in agriculture and for plaster. Previously it was also used locally in iron smelting, using charcoal from the nearby forests. Rare and very small fossils found in the limestone indicate that it is Early Ordovician in age (mid-Arenig, about 475 million years old).

Leave the quarry and walk round to the right on the path to the mast and viewpoint at the top of the hill. Carboniferous lavas of the Campsie

Figure 10.19 Aberfoyle: Old Red Sandstone boulder conglomerate, vertical, adjacent to Highland Boundary Fault.

Hills are visible, with Old Red Sandstone crags on the near horizon. To the north are the Dalradian mountains of Ben Lomond, Ben Venue and Ben Lui. The main Aberfoyle slate quarries can be seen nearby. This slate, which was worked for over 200 years, belongs to the Southern Highland Group of the Dalradian. It is a relatively good quality slate, and was used to roof many of the houses in Central Scotland.

Return to the quarry and take the broad, more gentle path through the forest and down the hill. At the edge of the track near the first crossroads is an exposure of the Achray Sandstone, a member of the Highland Border Complex. It is vertical, pale yellowish grey and has rusty joint faces (from the weathering-out of iron). At the bend by the next crossroads, the vertical dark crags seen through the trees are of grit. Some 300m farther down the track is a clearing where the waterfall comes into view. The waterfall is at a boundary between grit and slate. Continue downhill and just after the sharp bend in the stream alongside the path are some vertical bluish to greenish phyllites or slates, very fine-grained and with a crenulation cleavage surface, resembling fine corrugations. Carry on over the footbridge and take the path to the waterfall to examine the vertical green phyllite and much tougher reddish grit layers, previously mud and coarse angular sand respectively, laid down in deep water as turbidity-current sediments prior to folding and metamorphism. Return to the lodge.

Additional excursions to the Trossachs area can be found in Browne and Gillen (2015), Lawson and Weedon (1992), Stephenson *et al.* (2013), and Treagus (2009).

Selected references and further reading

General geology of the Scottish Highlands

Betterton, J., Craig, J., Mendum, J.R., Neller, R. & Tanner, J. (eds). 2019. *Aspects of the Life and Works of Archibald Geikie*. London, Geological Society Special Publications, No. 480. 406 pages. (Includes a discussion of Geikie's role in the 'Highland controversy').

Dryburgh, P.M., Ross, S.M. & Thompson, C.L. 2014. *Assynt: the Geologists' Mecca* (second edition). Edinburgh: Edinburgh Geological Society. 38 pages. (A booklet detailing the early history of research in Assynt, and the famous Highland geological controversy of 1860.) Available free online from Edinburgh Geological Society website.

Emeleus, C.H. & Bell, B.R. 2005. *British Regional Geology: The Palaeogene Volcanic Districts of Scotland*. Nottingham: British Geological Survey. 214 pages. (Fourth edition of the 'Tertiary volcanic districts'; includes Ardnamurchan, plus a coloured geological map of the Inner Hebrides.) Available online at BGS Earthwise.

Friend, P. 2012. *Scotland – Looking at the Natural Landscapes*. London: HarperCollins New Naturalist Library, vol. 119. 466 pages.

Gillen, C. 2013. *Geology and Landscapes of Scotland* (second edition). Edinburgh: Dunedin Academic Press. 246 pages.

Hunter, A. & Easterbrook, G. 2004. *The Geological History of the British Isles*. Milton Keynes: Open University. 143 pages. (Has a good, well-illustrated introduction to the evolution of the Scottish Highlands.)

Johnstone, G.S. & Mykura, W. 1989. *British Regional Geology: The Northern Highlands of Scotland* (fourth edition). 219 pages. London: British Geological Survey/Her Majesty's Stationery Office. Available free online from Earthwise, BGS.

Jones, K. & Blake, S. 2003. *Mountain Building in Scotland*. Milton Keynes: Open University. 116 pages. (Well-illustrated, authoritative and excellent account of the origin of Scotland's mountains.)

Kempe, N. & Wrightham, M. (eds) 2006. *Hostile Habitats: Scotland's Mountain Environment*. Edinburgh: Scottish Mountaineering Trust. 256 pages. (This hillwalker's guide to the landscape and wildlife of upland areas of Scotland has well-illustrated sections on geology, landforms and conservation.)

Kokelaar, B.P. & Moore, I.D. 2006. *Glencoe Caldera Volcano, Scotland*. Nottingham: British Geological Survey. 127 pages. (Classic areas of British geology, to accompany a 1:25,000 scale geological map; includes suggestions for excursions.) Available online from BGS.

Law, R.D., Butler, R.W.H., Holdsworth, R.E., Krabbendam, M. & Strachan, R. (eds) 2010. *Continental Tectonics and Mountain Building – the Legacy of Peach and Horne.* London: Geological Society, Special Publication No. 335. 880 pages. (Proceedings of a 2007 conference in Ullapool to celebrate the centenary of the publication of the Peach and Horne memoir of 1907; extensive reference list to over 100 years of research.)

McKirdy, A. 2015. *Set in Stone: the Geology and Landscapes of Scotland.* Edinburgh: Birlinn. 102 pages. (Well-illustrated short introduction to Scotland's geology and scenery; the author and publisher have a collection of regional booklets ('Landscapes in Stone'), similar in scope to the BGS/SNH series 'Landscape Fashioned by Geology' and freely available online.)

McKirdy, A., Gordon, J. & Crofts, R. 2007. *Land of Mountain and Flood: the Geology and Landforms of Scotland.* Edinburgh: Birlinn and Scottish Natural Heritage. 324 pages. (A full-colour, large-format book, introducing the geology of Scotland.)

Oldroyd, D.R. 1990. *The Highlands Controversy.* Chicago: University of Chicago Press. 438 pages. (Details the fascinating and often turbulent history of how geological knowledge was constructed in the nineteenth century through fieldwork in the Northwest Highlands).

Park, R.G. 2002. *The Lewisian Geology of Gairloch, NW Scotland.* Geological Society, London. Memoir No. 26. 88 pages. (Research memoir, with detailed full-colour 1:20,000 scale geological map).

Park, R.G. 2012. *Introducing Tectonics, Rock Structures and Mountain Belts.* Edinburgh: Dunedin Academic Press. 132 pages. (Includes chapters on ancient orogenic belts in Scotland, and detailed explanations of the mechanics of folding, shearing and thrust faulting; with an extensive glossary.)

Park, R.G. 2014. *The Making of Europe: A Geological History.* Edinburgh: Dunedin Academic Press. 164 pages. (With a useful summary of Precambrian and Caledonian evolution of Scotland.)

Park, R.G. 2017. *Mountains: the Origins of the Earth's Mountain Systems.* Edinburgh: Dunedin Academic Press. 256 pages. (Includes a chapter on the Caledonian mountains.)

Rider, M. and Harrison, P. 2019. *Hutton's Arse: 3 billion years of extraordinary geology in Scotland's Northern Highlands* (second edition). Edinburgh: Dunedin Academic Press. 240 pages.

Roberts, J.L. 1989. *The Macmillan Field Guide to Geological Structures.* London: Macmillan. (250 pages of colour illustrations showing a great variety of rock structures found in Scotland, including those in the Highland Geology Trail.)

Roberts, J.L. 1998. *The Highland Geology Trail.* Edinburgh: Luath Press. 107 pages. (A driver's route guide, anticlockwise from Inverness to Oban.)

Stewart, A.D. 2002. *The Later Proterozoic Torridonian rocks of Scotland: their Sedimentology, Geochemistry and Origin.* Geological Society, London, Memoir

No. 24. 125 pages. (Contains a black-and-white sketch map and field directory to Torridonian localities in Northwest Scotland.)

Stone, P. 2008. *Bedrock Geology UK North*. Nottingham: British Geological Survey. 88 pages. (Booklet to accompany fifth edition of geological map of Scotland, 1:625,000 scale, sold as a pack.) Booklet available online at BGS Earthwise.

Trewin, N.H. (ed.) 2002. *The Geology of Scotland* (fourth edition). London: Geological Society. 576 pages.

Upton, B. 2015. *Volcanoes and the Making of Scotland* (second edition). Edinburgh: Dunedin Academic Press. 248 pages. (Includes chapters on volcanic rocks of Ardnamurchan, Ben Nevis, Glen Coe and igneous rocks in the Lewisian and Caledonian belts.)

Williams, N. (ed.) 2010. *The Birth of Ben Nevis*. Fort William: Lochaber Geopark. 47 pages. (Introduction to the geology and landscape within the geopark; including Glen Coe, Glen Nevis, Ardnamurchan, the Great Glen and the Parallel Roads of Glen Roy; full colour maps and photographs.)

Woodcock, N.H. & Strachan, R.A. (eds) 2010. *Geological History of Britain and Ireland* (second edition). Oxford: Blackwell Science. 442 pages.

Geological excursion guides to the Scottish Highlands

Barber, A.J., Beach, A., Park, R.G., Tarney, J. & Stewart, A.D. 1978. *The Lewisian and Torridonian Rocks of North-West Scotland*. Geologists' Association Guide, No. 21. London: Geologists' Association. 99 pages.

Browne, M.A.E. & Gillen, C. (eds) 2015. *A Geological Excursion Guide to the Stirling and Perth Area*. Edinburgh: Edinburgh Geological Society in association with NMS Enterprises Limited. 231 pages. Available free online at BGS Earthwise.

Gannon, P. 2012. *Rock Trails: Scottish Highlands*. Caernarfon: Pesda Press. 252 pages. (18 hill walks with geological descriptions, maps and photographs.)

Goodenough, K.M. & Krabbendam, M. 2011. *A Geological Excursion Guide to the Northwest Highlands of Scotland*. Edinburgh: Edinburgh Geological Society and National Museums of Scotland Publishing. 216 pages. Available free online, at Earthwise, BGS.

Goodenough, K.M., Pickett, E., Krabbendam, M. & Bradwell, T. 2004. *Exploring the Landscape of Assynt*. Edinburgh: British Geological Survey. 53 pages. (Booklet plus 1:50,000 scale geological map. A walkers' guide, covering Knockan, Inchnadamph Bone Caves, Traligill Caves, Achmelvich, Clachtoll, Stac Pollaidh, Suilven, Quinag, Conival and Ben More Assynt, Loch Glencoul.)

Gribble, C.D. 1976. *Ardnamurchan: a Guide to Geological Excursions*. Edinburgh: Edinburgh Geological Society. 122 pages. (With a 1:20,000 scale colour geological map.) Available online at BGS Earthwise.

Hambrey, M.J. and others, 1991. *The Late Precambrian Geology of the Scottish Highlands and Islands*. Geologists' Association Guide, No. 44. London: Geologists' Association. 130 pages.

Lawson, J.D. & Weedon, D.S. 1992. *Geological Excursions around Glasgow and Girvan*. Glasgow: Geological Society of Glasgow. 495 pages. Available online at BGS

Earthwise.

Pattison, D.R.M. & Harte, B. 2001. *The Ballachulish Igneous Complex and Aureole: A Field Guide*. Edinburgh: Edinburgh Geological Society. 150 pages. (With full colour 1:40,000 scale geological map and excellent field photographs of Dalradian metamorphic rocks.) Available online at BGS Earthwise.

Roberts, J.L. & Treagus, J. 1977. The Dalradian rocks of the South-west Highlands. *Scottish Journal of Geology*, vol. **13**, part 2, pages 85–184. (Introduction and 7 field excursions.) Available online at BGS Earthwise.

Storey, C. 2008. The Glenelg–Attadale Inlier, NW Scotland. Introduction, review and field guide. *Scottish Journal of Geology*, vol. **44**, pages 17–34. (Detailed maps and photographs in full colour.)

Strachan, R., Alsop, I., Friend, C. & Miller, S. (eds) 2010. *A geological excursion Guide to the Moine geology of the Northern Highlands of Scotland* (second edition). Edinburgh: Edinburgh Geological Society and National Museums of Scotland Publishing. 298 pages. Available online at BGS Earthwise.

Treagus, J. 2009. *The Dalradian of Scotland*. Geologists' Association Guide No. 67. London: Geologists' Association. 202 pages. (22 excursions to the Southwest Highlands, Central Highlands and Banffshire coast.)

Trewin, N.H. & Hurst, A. (eds) 2009. *Excursion Guide to the Geology of East Sutherland and Caithness* (second edition). Edinburgh: Dunedin Academic Press. 183 pages.

Williams, N. 2016. *Exploring the Landscape of Ben Nevis and Glen Nevis*. Fort William: Nevis Landscape Partnership. 64 pages. (A walkers' guide to the rocks and landscapes of Ben Nevis and Glen Nevis; 8 short excursions; accompanied by 1:25,000 scale geological sketch map, and many photographs.)

Geological Conservation Review Series

Gordon, J.E. & Sutherland, D.G. (eds) 1993. *Quaternary of Scotland*. Geological Conservation Review Series, No. 6. London: Chapman & Hall. 695 pages. (For more recent research in this area, see Ballantyne, C.K. & Small, D. 2018, and Sissons, 2017, below.)

Mendum, J., Barber, A.J., Butler, R.W.H. and others. 2009. *Lewisian, Torridonian and Moine Rocks of Scotland*. Geological Conservation Review Series, No. 34. Peterborough: Joint Nature Conservation Committee. 721 pages. (Includes many black-and-white sketch maps and photographs, and detailed descriptions of key geological sites in the Highlands written by a large panel of experts.)

Stephenson, D., Leslie, A.G., Mendum, J.R., Tanner, P.W.G. & Treagus, J.G. (eds) 2013. Dalradian Rocks of Scotland. Geological Conservation Review Series. *Proceedings of the Geologists' Association*, vol. **124**, issues 1–2, pages 1–410.

Stephenson, D., Bevins, R.E., Millward, D., Stone, P., Parsons, I., Highton, A.J. & Wadsworth, W.J. 2000. *Caledonian Igneous Rocks of Great Britain*. Geological Conservation Review Series, No. 17. Peterborough: Joint Nature Conservation Committee. 648 pages. Available online at BGS Earthwise.

Academic research papers

Anderton, R. 1985. Sedimentation and tectonics in the Scottish Dalradian. *Scottish Journal of Geology*, vol. **21**, pages 407–436.

Amor, K. *et al.* 2019. The Mesoproterozoic Stac Fada proximal ejecta blanket, NW Scotland: constraints on crater location from field observations, anisotropy of magnetic susceptibility, petrography and geochemistry. *Journal of the Geological Society of London*, vol. **176**.

Baker, T.R., Prave, A.R. & Spencer, C.J. 2019. 1.99 Ga mafic magmatism in the Rona terrane of the Lewisian Gneiss Complex in Scotland. *Precambrian Research*, vol. **329**, pages 224–231.

Ballantyne, C.K. & Small, D. 2018. The last Scottish ice sheet. *Earth and Environmental Science Transactions of the Royal Society of Edinburgh*, vol. **109**, pages 1–39.

Barooah, B.C. & Bowes, D.R. 2009. Multi-episodic modifications of a high-grade terrane near Scourie and its significance in elucidating the history of the Lewisian Complex. *Scottish Journal of Geology*, vol. **45**, pages 19–42.

Batchelor, R.A. 2008. Lamprophyric magmatism in the Sleat Group (Torridonian), Kyle of Lochalsh: Torridonian sill or ash-flow tuff, or Permo-Carboniferous intrusion? *Scottish Journal of Geology*, vol. **44**, pages 35–41.

Beach, A., Coward, M.P. & Graham, R.H. 1974. An interpretation of the structural evolution of the Laxford front. *Scottish Journal of Geology*, vol. **9**, pages 297–308.

Bird, A., Cutts, K., Strachan, R., Thirlwall, M.F. & Hand, M. 2018. First evidence of Renlandian (*c.* 950–940Ma) orogeny in mainland Scotland: Implications for the status of the Moine Supergroup and circum-North Atlantic correlations. *Precambrian Research*, vol. **305**, pages 283–294.

Bluck, B.J. 2015. The Lower Old Red Sandstone at Balmaha–Aberfoyle and its bearing on the nature of the Highland Boundary and Gualann faults. *Scottish Journal of Geology*, vol. **51**, pages 165–176.

Brasier, A.T., Culwick, T. Battison, L. Callow, R. H.T. & Brasier, M.D. 2017. Evaluating evidence from the Torridonian Supergroup (Scotland, UK) for eukaryotic life on land in the Proterozoic. In: Brasier, A.T., McIlroy, D. & McLoughlin, N. (eds) *Earth System Evolution and Early Life: a Celebration of the Work of Martin Brasier*. Geological Society, London, Special Publications, No. 448, pages 121–144.

Burchard, S., Troll, V.R., Mathieu, L., Emeleus, H.C. & Donaldson, C.H. 2013. Ardnamurchan 3D cone-sheet architecture explained by a single elongate magma chamber. *Scientific Reports* **3**, 2891. Nature Publications, available online: DOI:10.1038/srep02891 (2013).

Butler, R.W.H. 2010. *The geological structure of the North-West Highlands of Scotland revisited: Peach et al. 100 years on*. London: Geological Society, Special Publications, vol. **335**, pages 7–27, In: Law *et al.* (see above).

Butler, R.W.H., Matthews, S.J. & Morgan, R.K. 2007. Structural evolution of the Achnashellach Culmination, southern Moine Thrust Belt; testing the duplex model. In: Ries, A.C., Butler, R.W.H. & Graham, R.H. (eds) *Deformation of the Continental Crust: the Legacy of Mike Coward*. Geological Society, London, Special Publications, vol. **272**, pages 103–120.

Cain, T., Leslie, G., Clarke, S., Kelly, M. & Krabbendam, M. 2016. Evidence for pre-Caledonian discontinuities in the Achnashellach Culmination, Moine Thrust Zone: the importance of a pre-thrust template in influencing fold-and-thrust belt development. *Scottish Journal of Geology*, vol. **52**, pages 103–109.

Cawood, P.A., Strachan, R.A. and others. 2015. Neoproterozoic to early Paleozoic extensional and compressional history of East Laurentian margin sequences: the Moine Supergroup, Scottish Caledonides. *Geological Society of America Bulletin*, vol. **127**, pages 349–371.

Chew, D.M. & Strachan, R.A. 2013. The Laurentian Caledonides of Scotland and Ireland. In: Corfu, F., Gasser, D. & Chew, D.M. (eds) *New Perspectives on the Caledonides of Scandinavia and Related Areas*. Geological Society, London, Special Publications, vol. **390**, pages 45–91.

Cresswell, D. 1972. The structural development of the Lewisian rocks on the north shore of Loch Torridon, Ross-shire. *Scottish Journal of Geology*, vol. **8**, pages 293–308.

Dash, B. & Bowes, D.R. 2014. Lewisian Complex of Strath Dionard–Rhiconich and its significance in the early history of the NW Highlands of Scotland. *Scottish Journal of Geology*, vol. **50**, pages 27–48.

Davies, J.H.F.L. & Heaman, L.M. 2014. New U–Pb baddeleyite and zircon ages for the Scourie dyke swarm: a long-lived large igneous province with implications for the Paleoproterozoic evolution of NW Scotland. *Precambrian Research*, vol. **249**, pages 180–198.

Findlay, D. & Bowes, D.R. 2017. Structural framework of the gneiss–amphibolite-pegmatite assemblage of the Lewisian Complex south of Durness, NW Highlands. *Scottish Journal of Geology*, vol. **53**, pages 13–28.

Goodenough, K.M., Crowley, Q.G., Krabbendam, M. & Parry, S.F. 2013. New U–Pb age constraints for the Laxford Shear Zone, NW Scotland: Evidence for tectono-magmatic processes associated with the formation of a Paleoproterozoic supercontinent. *Precambrian Research*, vol. **233**, pages 1–19.

Guice, G.L., McDonald, I., Hughes, H.S.R., MacDonald, J.M., Blenkinsop, T.G., Goodenough, K.M., Faithfull, J.W. & Gooday, R.J. 2018. Re-evaluating ambiguous age relationships in Archean cratons: implications for the origin of ultramafic–mafic complexes in the Lewisian Gneiss Complex. *Precambrian Research*, vol. **311**, pages 136–156.

Hardman, K. 2019. Cracking Canisp: deep void evolution during ancient earthquakes. *Geoscientist*, Geological Society of London, vol. **29**, no. 1, pages 10–16. (A brief and well-illustrated history of the long-lived Canisp shear zone at Achmelvich.)

Henderson, W.G., Tanner, P.W.G. & Strachan, R.A. 2009. The Highland Border Ophiolite of Scotland. *Scottish Journal of Geology*, vol. **45**, pages 13–18.

Hughes, H.S.R., Goodenough K.M., Walters, A.S., McCormac, M., Gunn, G.A. & Lacinska, A. 2013. The structure and petrology of the Cnoc nan Cuilean Intrusion, Loch Loyal Syenite Complex, NW Scotland. *Geological Magazine*, vol. **150**, pages 783–800.

Hutton, D.H.W. 1988. Igneous emplacement in a shear-zone termination: the biotite granite at Strontian, Scotland. *Geological Society of America Bulletin*, vol. **100**, pages 1392–99.

Kinny, P.D., Friend, C.R.L. & Love, G.J. 2005. Proposal for a terrane-based nomenclature for the Lewisian Gneiss Complex of NW Scotland. *Journal of the Geological Society of London*, vol. **162**, pages 175–186.

Krabbendam, M., Bonsor, H., Horstwood, S.A. & Rivers, T. 2017. Tracking the evolution of the Grenvillian foreland basin: constraints from sedimentology and detrital zircon and rutile in the Sleat and Torridon groups, Scotland. *Precambrian Research*, vol. **295**, pages 67–89.

Krabbendam, M., Leslie, A.G. & Goodenough, K.M. 2014. Structure and stratigraphy of the Morar Group in Knoydart, NW Highlands: implications for the history of the Moine Nappe and statigraphic links between the Moine and Torridonian successions. *Scottish Journal of Geology*, vol. **50**, pages 125–144.

Krabbendam, M., Ramsay, J. G. Leslie, A. G., Tanner, P. W. G., Dietrich, D. & Goodenough, K. M. 2018. Caledonian and Knoydartian overprinting of a Grenvillian inlier and the enclosing Morar Group rocks: structural evolution of the Precambrian Proto-Moine Nappe, Glenelg, NW Scotland. *Scottish Journal of Geology*, vol. **54**, pages 1–23. (Detailed maps, cross-sections and field photographs in full colour.)

Leslie, A.G., Smith, M. & Soper, N.J. 2008. Laurentian margin evolution and the Caledonian Orogeny – a template for Scotland and East Greenland. In: Higgins, A.K., Gilotti, J.A., & Smith, M.P. (eds) *The Greenland Caledonides – Evolution of the Northeast margin of Laurentia*. Geological Society of America, Memoir No. 202, pages 307–343.

Muir, R.J. & Vaughan, A.P.M. 2017. Digital mapping and three-dimensional model building of the Ben Nevis Igneous Complex, Southwest Highlands, Scotland: new insights into the emplacement and preservation of postorogenic magmatism. *The Journal of Geology*, vol. **125**, pages 607–636.

O'Driscoll, B. 2007. The Centre 3 layered gabbro intrusion, Ardnamurchan, NW Scotland. *Geological Magazine*, vol. **144**, pages 897–908.

Park, R.G. 2005. The Lewisian terrane model: a review. *Scottish Journal of Geology*, vol. **41**, pages 105–118.

Park, R.G. 2010. Structure and evolution of the Lewisian Gairloch shear zone: variable movement directions in a strike-slip regime. *Scottish Journal of Geology*, vol. **46**, pages 31–44.

Park, R.G., Tarney, J. & Connelly, J.N. 2001. The Loch Maree Group: Palaeoproterozoic subduction-accretion complex in the Lewisian of NW Scotland. *Precambrian Research*, vol. **105**, pages 205–226.

Rollinson, H. & Gravestock, P. 2012. The trace element geochemistry of clinopyroxenes from pyroxenites in the Lewisian of NW Scotland: insights into light rare earth element mobility during granulite facies metamorphism. *Contributions to Mineralogy and Petrology*, vol. **168**, pages 319–335.

Searle, M., Cornish, S.B., Heard, A., Charles, J.-H. & Branch, J. 2019. Structure of the

Northern Moine thrust zone, Loch Eriboll, Scottish Caledonides. *Tectonophysics*, vol. **752**, pages 35–51.

Simms, M.J. 2015. The Stac Fada impact ejecta deposit and the Lairg Gravity Low: evidence for a buried Precambrian impact crater in Scotland? *Proceedings of the Geologists' Association*, vol. **126**, pages 742–761.

Simms, M.J. & Ernstson, K. 2019. A reassessment of the proposed 'Lairg Impact Structure' and its potential implications for the deep structure of northern Scotland. *Journal of the Geological Society of London*, vol. **176**.

Sissons, B. 2017. The lateglacial lakes of glens Roy, Spean and vicinity (Lochaber district, Scottish Highlands). *Proceedings of the Geologists' Association*, vol. **128**, pages 32–41.

Sissons, B. 2017. The varve-related ice-dammed lake events in Glen Roy and vicinity: a new interpretation. *Proceedings of the Geologists' Association*, vol. **128**, pages 146–150.

Smith, D.E., Barlow, N.L.M., Bradley, S.L., Firth, C.R., Hall, A.M., Jordan, J.T. & Long, D. 2017. Quaternary sea level change in Scotland. *Earth and Environmental Science Transactions of the Royal Society of Edinburgh*, vol. **108**, pages 1–38.

Tanner, P.W.G. 2012. The giant quartz-breccia veins of the Tyndrum–Dalmally area, Grampian Highlands, Scotland: their geometry, origin and relationship to the Cononish gold–silver deposit. *Earth and Environmental Science Transactions of the Royal Society of Edinburgh*, vol. **103**, pages 51–76.

Tanner, P.W.G. & Sutherland, S. 2007. The Highland Border Complex, Scotland: a paradox resolved. *Journal of the Geological Society*, London, vol. **164**, pages 111 -116.

Tanner, P.W.G. & Thomas, P.R. 2010. Major nappe-like D2 folds in the Dalradian rocks of the Beinn Udlaidh area, Central Highlands, Scotland. *Earth and Environmental Science Transactions of the Royal Society of Edinburgh*, vol. **100**, pages 371–389.

Tarney, J. 1963. Assynt dykes and their metamorphism. *Nature*, vol. **199**, pages 672–4.

Wacey, D. and others. 2017. Contrasting microfossil preservation and lake chemistries within the 1200–1000Ma Torridonian Supergroup of NW Scotland. In: Brasier, A.T., McIlroy, D. & McLoughlin, N. (eds) *Earth System Evolution and Early Life: a Celebration of the Work of Martin Brasier*. Geological Society, London, Special Publications, No. 448, pages 105–119.

Walsh, J.A. 2000. *Scottish Slate Quarries*. Technical Advice Note 21. Edinburgh: Historic Scotland. 113 pages. (Includes detailed descriptions of Ballachulish, Easdale and Aberfoyle quarries.) BGS website (Earthwise) has an extensive set of photographs illustrating slates from all the Scottish quarries.

Watkins, H., Butler, R.W.H., Bond, C.E. & Healy, D. 2015. Influence of structural position on fracture networks in the Torridon Group, Achnashellach fold and thrust belt, NW Scotland. *Journal of Structural Geology*, vol. **74**, pages 64–80.

Wheeler, J. 2007. A major high-strain zone in the Lewisian Complex in the Loch Torridon area, NW Scotland: insights into deep crustal deformation. In: Ries, A.C., Butler, R.W.H. & Graham R.H. (eds) *Deformation of the Continental Crust: The Legacy of Mike Coward*. London: Geological Society London, Special Publications, No. 272, pages 27–45.

Selected references and further reading

Wheeler, J., Park, R.G., Rollinson, H.R. & Beach, A. 2010. The Lewisian Complex: insights into deep crustal evolution. In: Law, R.D., Butler, R.W.H., Holdsworth, R.E., Krabbendam, M. & Strachan, R.A. (eds) *Continental tectonics and Mountain building: the legacy of Peach and Horne*. Geological Society, London, Special Publications, No. 335, pages 51–79.

Appendix

Websites

Geological societies

These societies organize summer field excursions, winter lectures, and publish geological excursion guides:

Aberdeen geological society: www.aberdeengeolsoc.co.uk

Edinburgh geological society: www.edinburghgeolsoc.org

Glasgow geological society: www.geologyglasgow.org.uk

Highland geological society: www.spanglefish.com/highlandgeologicalsociety

London geological society (bookshop sells field guides online): www.geolsoc.org.uk

Open University geological society: www.ougs.org

Other useful sites

British Geological Survey (BGS): www.bgs.ac.uk (Earthwise portal has field guides to download for free: http://earthwise.bgs.ac.uk/index.php/Category:Scotland.) The map portal gives access to all published geological maps: http://mapapps.bgs.ac.uk/geologyofbritain/home.html. Sheet memoirs can also be accessed from the map list. Detailed geology can be viewed on the Onshore Geoindex: https://www.bgs.ac.uk/geoindex/. The 3D version can be tilted and rotated: http://mapapps.bgs.ac.uk/geologyofbritain3d/index.html). BGS online shop (for maps, guides and books): www.bgs.ac.uk/shop. A 3D interactive model and video of the Assynt Culmination of the Moine Thrust zone can be found here: https://www.bgs.ac.uk/research/ukgeology/scotland/assyntCulmination.html

Geograph (maps, photographs with detailed captions, footpaths and walks, for every grid square in the country): https://www.geograph.org.uk.

Joint Nature Conservation Committee (JNCC) site for geological conservation (Earth heritage and geological conservation review (GCR) sites): www.jncc.gov.uk. Information on GCR sites (Lewisian, Torridonian & Moine; Dalradian; Caledonian igneous rocks; Quaternary of Scotland; Minerals of Scotland) can be found there.

Leeds University, Assynt geology website: https://www.see.leeds.ac.uk/structure/. The structure pages contain much useful information on folds, faults, thrusting, shear zones, cleavage formation and other topics relevant to the geology of the Highlands.

Leeds University, Moine Thrust pages: www.see.leeds.ac.uk/structure/mtb/index.

htm. This site is very extensive, with detailed notes, explanations, photographs, maps and cross-sections dealing with the structure of the Moine Thrust zone in Northwest Scotland.

Loch Lomond national park: www.lochlomond-trossachs.org

Lochaber Geopark: www.lochabergeopark.org.uk. The map page shows the detailed geology of the geopark area: https://lochabergeopark.org.uk/aboutus/geological-maps/. There is a link to a video on the formation of Ben Nevis: https://www.youtube.com/watch?v=M4QYlkUAI1o, but note that more-recent remapping has resulted in a new interpretation (see Muir & Vaughan, 2017).

National Trust for Scotland: https://www.nts.org.uk. The trust is responsible for several scenic areas in the Highlands, including Torridon, Kintail and Glencoe.

Northwest Highlands Geopark: www.northwest-highlands-geopark.org.uk; see also the Wester Ross Coigach Geotrail leaflet with a series of short excursions: https://coigach.com/see-and-do/the-coigach-geotrail/.

Oxford University, Rocks of Northwest Scotland: www.earth.ox.ac.uk/~oesis/nws/nws-home.html (includes an extensive atlas of field photographs, hand specimens and microscope thin sections).

Scottish Geology: www.scottishgeology.com (this has links to the 51 Best Places to see geology in Scotland; 10 of the sites are in this book).

Scottish Mountaineering Club: www.smc.org.uk

Scottish Natural Heritage (SNH): https://www.nature.scot (the 'Landscape fashioned by geology' (published jointly with BGS) series of booklets can be downloaded from here).

Undiscovered Scotland: www.undiscoveredscotland.co.uk (for accommodation, walks, transport links and maps).

Walkhighlands (for hillwalkers): www.walkhighlands.co.uk. This site has detailed route maps and illustrated walking guides for most of the localities in this book.

Wester Ross geology: Jeremy Fenton's site has a good section on geology, and a 40 page coloured leaflet to download: http://www.jeremyfenton.scot/Booklet%20 Geology%20lo.pdf

Accommodation

Visit Scotland: www.visitscotland.com

www.welcome-scotland.com (also includes information on travel and outdoor activities)

To search by map: www.roomfinderscotland.co.uk

Scottish Youth Hostels Association: www.syha.org.uk

Independent hostels: www.hostel-scotland.co.uk

Transport

General information with links: www.visithighlands.com/travel

Caledonian–MacBrayne (Calmac) ferries: www.calmac.co.uk

Corran ferry: www.lochabertransport.org.uk

Glenelg–Kylerhea ferry: www.skyeferry.co.uk

Ellenabeich–Easdale ferry: https://www.argyll-bute.gov.uk/
    ellenabeich-easdale-ferry-timetable

Garvellachs ferry: https://www.seafari.co.uk/oban/

Glossary

A

accretion   growth of continents by collision; or movement together of blocks of rock.

acidic   igneous rock with over 65% silica, e.g. granite, rhyolite; same as **felsic**.

agmatite   black hornblende gneiss blocks surrounded by white feldspar-rich veins and patches; a type of migmatite; occurs in Central region of Lewisian Gneiss Complex.

alkaline igneous rocks   igneous rocks with high potassium and sodium contents relative to silica; they are considered rather unusual rock types and frequently contain rare-earth minerals; an example is syenite (in contrast, rocks such as granite and rhyolite do not contain alkaline minerals).

alluvial deposits   river sediments.

alluvial fan   river deposits laid down at the mouth of a valley that opens out onto a flat plain at the edge of a steep mountain slope; fan shape spreads outwards and sediments become finer and thin away from the valley mouth.

amphibole   a large family of silicate minerals made of calcium, iron and magnesium (**'ferromagnesian'**); dark green to black; elongate crystals; a common member of the family is hornblende.

amphibolite   hornblende–feldspar–biotite gneiss, often with garnet; formed from high-grade metamorphism of basic lavas or dolerite sills and dykes.

andesite   igneous rock, lava of intermediate composition; named after the Andes mountains.

anorthosite   coarse-grained layered igneous rock containing mostly anorthite, a calcium-rich plagioclase feldspar.

anticline   fold shape with limbs dipping away from axis; oldest rocks in centre or core of fold.

Appin Group   part of Dalradian Supergroup, above Grampian and below Argyll groups; divided into Lochaber, Ballachulish and Blair Atholl subgroups.

Applecross Formation   part of Torridon Group, above Diabaig and below Aultbea formations; consists of coarse red sandstones.

Archaean (Archean)   Precambrian rocks aged between 4000 and 2500 million years old.

Argyll Group   part of Dalradian Supergroup, above Appin and below Southern Highland groups; divided into Islay, Easdale, Crinan and Tayvallich subgroups.

arkose   type of coarse sandstone with over 20% feldspar fragments; red or brown in colour; e.g. Torridonian sandstone; preferred term is now feldspathic sandstone.

ash   fine volcanic material from explosive eruptions.

Assynt culmination   eastward indentation of Moine Thrust zone caused by igneous intrusions and stacking-up of thrust sheets and nappes.

augen gneiss   gneiss with large eye-shaped feldspar crystals surrounded by mica; a type of deformed granite.

aureole   zone of country rock around an igneous intrusion, affected by heat and fluids, causing **contact metamorphism**.

Avalonia   Precambrian continent containing southern Britain, Ireland and western Europe; collided with Laurentia and Baltica to create the Caledonian mountain belt as the Iapetus Ocean closed 420 million years ago.

B

Badcallian   Precambrian (2900 million years ago) folding and metamorphism of the Lewisian rocks around Scourie; previously called early Scourian.

Badenoch Group   sequence of high-grade metamorphic rocks forming the basement to the Dalradian in the northern Grampian Highlands; folding and metamorphism at 800 million years does not affect the overlying Grampian Group of the Dalradian.

Baltica   Precambrian continent containing Scandinavia; collided with Laurentia and Avalonia to create the Caledonian mountain belt.

banded ironstone (banded iron formation)   Precambrian sedimentary rock made of iron ore and chert (silica).

banding   layering of different minerals in metamorphic rocks, usually gneisses.

basalt   dark, fine-grained basic volcanic lava; contains olivine, pyroxene, calcium-rich plagioclase feldspar, iron ore.

basement   older rocks on which sediments have been deposited; usually refers to a Precambrian **craton**, e.g. Lewisian gneiss basement of the Northwest Highlands.

basic igneous rock   rock with 44–52% silica; rich in ferromagnesian minerals (olivine, pyroxene) and calcium-rich plagioclase feldspar, no free quartz crystals; e.g. basalt, dolerite, gabbro; same as **mafic**.

basin   broad area of slowly subsiding continental crust into which sediment has been transported by rivers; some basins are rift valleys or graben structures when bounded by two parallel normal faults.

biotite   black or dark brown flaky iron-rich mica (silicate mineral) with sheet structure.

boudinage   black-pudding-shaped structure in deformed rocks, due to thinning and stretching of layers of different composition and strength.

breccia   rock consisting of broken fragments; formed in volcanic explosions, or along fault lines, or as talus (scree) formed by collapse of rock slopes.

brittle   property of a rock that breaks by fracturing.

C

calcite   calcium carbonate; mineral found in limestone and marble; white or colourless, soft, dissolves in weak acids.

Glossary

caldera large, circular volcanic structure with steep walls, usually marking a ring fault, and a flat, depressed downfaulted floor containing volcanic vents.

Caledonian orogeny mountain-building episode 500–400 million years ago that created the mountain chain in Ireland, Scotland, Wales, western Norway and Greenland. Early events are referred to as Grampian, later as Scandian.

Cambrian Period geological time period, from 542 to 488 million years ago; above the Precambrian and below the Ordovician; named from Cambria, Roman name for Wales.

Cambrian Quarzite white rock made of quartz fragments cemented by silica, at base of Cambrian sequence in Northwest Highlands, west of Moine Thrust; forms prominent scree slopes on mountain tops.

Carboniferous Period geological period from 359 to 299 million years ago, above Devonian and below Permian. A time of tropical lagoons, large sandy rivers and coal swamps. Rocks of this age are very rare in the Highlands; small exposures exist near Lochaline.

cauldron subsidence collapse of a cylindrical volcanic crater and subsurface igneous intrusion into deeper levels of the crust, with the formation of a ring fracture; this is followed by upwelling of magma to form a ring intrusion near the surface.

Cenozoic Era division of the geological column, from 66 million years ago to the present, above the Cretaceous Period; includes the Tertiary (an informal but still common term comprising Palaeogene and Neogene) and Quaternary 'periods'; follows the **Mesozoic Era**; from Greek, meaning 'new life'. During the Cenozoic there was widespread volcanic activity on the west coast, followed by extensive deep sub-tropical weathering and erosion preceding the various Quaternary ice ages.

chalcopyrite copper–iron sulphide, an important copper ore mineral.

chert hard, compact form of silica found in chalk; formed of silica-rich organisms; similar to flint and jasper.

chilled margin zone of fine-grained rock at edge of an igneous body, formed due to rapid cooling.

chlorite green, flaky silicate mineral with sheet structure, related to mica; found in low-grade metamorphic rocks, slate and phyllite.

cleavage ability of a mineral or rock to break along a plane of weakness.

clint and gryke ridge and hollow pattern found in limestone country (karst), due to solution of carbonate by rain water.

collision zone boundary between two tectonic plates; continent–continent collision produces a mountain chain; ocean–ocean collision produces a volcanic island arc and trench (subduction zone); ocean–continent collision produces a trench and mountain chain with earthquakes and volcanoes.

columnar jointing closely spaced cooling joints in igneous rocks, e.g. lavas, sills and dykes.

concretion nodule of hard material with no internal structure, found in sedimentary rocks, e.g. chert in chalk, ironstone in sandstone, or limestone in sandstone.

cone sheet   igneous intrusion, circular at the top and tapering downwards to a point; cone sheets may be stacked one inside the other.

conformable   continuity of beds of sediment, laid down in sequence without time gaps or breaks in sedimentation.

conglomerate   sedimentary rock made of large and small rounded fragments.

constructive plate boundary   zone of oceanic crust where new material is added by injection of lava at mid-ocean ridge.

contact aureole   zone of thermal metamorphism around an igneous intrusion.

contact metamorphism   heat-affected zone of country rock surrounding an igneous intrusion; produces a **contact aureole**.

continental crust   top part of the continental lithosphere above the Moho; average 35km thick, up to 80km in mountain chains such as the Himalayas.

cordierite   magnesium–iron–aluminium silicate mineral found in some high-grade metamorphic rocks.

correlation   comparing sedimentary rock sequences, usually on the basis of their fossil content, and may be combined with rock type and environment.

corrie   glacial landform on a mountainside, caused by ice dislodging large, steep-walled segment of bedrock.

country rocks   rocks around an igneous intrusion or mineral vein.

cover rocks   sedimentary rocks deposited on older basement rocks.

craton   large stable area of ancient continental crust unaffected by earthquakes or volcanic activity since the Precambrian; also known as shield (Greek: strong).

crenulation   tiny folds like corrugations with parallel axes; common in phyllites and fine-grained schists.

Cretaceous Period   geological period from 146 to 66 million years ago, above Jurassic and below the Cenozoic Era. A time of warm chalk seas. Rocks of this age are almost absent from the Highlands, except for small patches in the Inner Hebrides.

cross bedding   structure in sedimentary rocks caused by currents depositing sand at different angles on sloping surfaces; also known as cross stratification.

crust   outermost layer of the Earth; continental crust is 35km thick on average, and up to 4 billion years old; oceanic crust is less than 10km thick and made of basalt, up to 180 million years old; lies above the Moho, with the mantle beneath.

D

Dalradian   sequence of late Precambrian to early Ordovician sedimentary and volcanic rocks in Scotland and Ireland, approximately 800–480 million years old, folded and metamorphosed during the Caledonian orogeny; found between the Great Glen Fault and the Highland Boundary Fault; also known as the Dalradian Supergroup.

destructive plate boundary   zone of the Earth's crust where one plate is consumed beneath another by subduction along an ocean trench; such boundaries are marked by earthquakes and volcanoes.

detrital zircons   zircon crystals found in sedimentary or metasedimentary rocks; can provide important clues to the age, origin and source of these rocks.

Devonian Period geological period from 416 to 359 million years ago, above Silurian and below Carboniferous. Old Red Sandstone rocks are mostly of this age. Found as small patches of red conglomerate around the north and west coasts, and lavas in the Lorn Plateau near Oban.

dextral rightwards-directed movement, e.g. of a fault.

diagenesis physical and chemical changes that take place in a loose sediment after deposition and during burial and lithification.

dip maximum inclination of a plane (e.g. a bed or a fault), at right angles to the strike.

displacement distance of physical movement.

dolerite medium-grained basic igneous rock, found in dykes and sills.

dolomite the mineral calcium–magnesium carbonate; also a form of limestone made of that mineral.

dolostone limestone made of dolomite.

downthrow vertical displacement of one side of a fault.

drift unsorted loose material plastered over the ground surface when ice sheets melt.

drumlin smooth, rounded hill of glacial sediment, formed by moving ice.

ductile deformation of rock by thinning, stretching and flow (opposite of brittle).

dune bedding form of cross bedding found in sand dunes; common in the New Red Sandstone (Permian to Triassic) of Scotland.

duplex deformation structure in which beds are pushed up at a steep angle between a sole thrust and a roof thrust; found in Moine Thrust Zone of Northwest Highlands.

Durness Group carbonate rocks lying above Cambrian quartzites, west of the Moine Thrust; mostly Ordovician in age.

dyke igneous intrusion in the form of a vertical sheet; may have formed a feeder to surface lava flow.

dyke swarm closely spaced set of vertical dykes radiating out from an igneous centre.

dynamic metamorphism process due to differential stress on rocks, creating new structures and textures due to shearing, e.g. mylonite.

E

eclogite high-pressure metamorphic rock formed from basalt slab taken down a subduction zone (to 50–60km); main minerals are red, glassy garnet and bright green pyroxene (Greek: specially chosen, on account of the unusual mineral composition).

epidote a distinctive bright green silicate mineral found in metamorphic rocks, often as streaks, patches and veins; complex water-rich structure, with calcium, aluminium and iron.

erratic boulder carried by moving ice and dropped far from its source.

exhumed topography buried ancient landscape exposed by erosion of overlying rocks; e.g. Torridonian on top of Lewisian in Northwest Highlands.

extension stretching and thinning of the crust, may lead to normal faulting and creation of sedimentary basins.

extrusive igneous lava poured out at the surface.

F

facies   total set of characteristics in a rock; in sedimentary rocks this will include grain composition and size, sedimentary structures, etc. and can be used to interpret past environment of deposition, e.g. river bed, delta, desert floor, shallow sea, lake or lagoon; in metamorphic rocks the facies is a combination of pressure and temperature conditions that produced a particular set of minerals under those conditions in a specific rock.

fan   a slope of rock detritus that widens down the slope and spreads out like a fan.

fault   break in rocks where there has been physical movement (displacement) between two blocks.

fault breccia   broken, angular fragments of wall rock, surrounded by crushed rock, in a fault zone.

fault displacement   relative movement of one block of rock against another.

feldspars   silicate minerals, with three-dimensional lattice structure; commonest minerals in the crust; plagioclase feldspars contain calcium and sodium; orthoclase feldspar contains potassium.

felsic   igneous rock containing light-coloured feldspar and quartz; contrasts with mafic; acid (acidic) is also used.

ferromagnesian minerals   silicate minerals containing iron and magnesium; e.g. olivine, pyroxene (augite), amphibole (hornblende).

fissile   describes a fine-grained rock that splits easily along bedding planes, e.g. shale (sedimentary), or along cleavages, e.g. slate (metamorphic).

Flat Belt   part of the inverted limb of the Tay Nappe fold, where the beds are flat-lying; this structure dominates the south-eastern Grampians.

flood basalt (plateau basalt)   basalt lava flows covering an extensive area, and usually forming stepped or **trap** landscape feature.

floodplain   low part of a river valley, which is liable to flooding; formed of alluvium deposited by meandering river.

flow banding   structure formed in igneous rock by lava being streaked out during flow.

fluvial sediments   sediments transported by rivers.

fold limb   part of a fold lying between fold hinges.

foliation   planar fabric in metamorphic rocks, usually used to describe schist or gneiss.

foreland   stable part of the continental crust in front of a mountain chain; folded rocks are frequently thrust over the foreland.

G

gabbro   a coarse-grained basic igneous rock containing olivine, pyroxene and plagioclase (calcium-rich feldspar); usually black speckled with cream or white.

galena   lead sulphide, PbS, that forms shiny bright grey cubes; very dense, an ore of lead.

garnet   a cubic silicate mineral (of iron, magnesium, aluminium), usually deep red in colour, found in metamorphic rocks that formed at medium to high pressure, e.g. schist or gneiss.

glacial erratics   boulders of rock carried by ice and deposited far from their source; useful for deducing ice-movement direction.

Glenfinnan Group   middle unit of Moine, 920–885 million years old; 3km thick unit of schist ('pelites'); lies above the Sgùrr Beag Thrust and below the Loch Eil Group.

gneiss   very coarse-grained, banded, high-grade metamorphic rock with foliation often marked by biotite mica.

Gondwana (Gondwanaland)   supercontinent formed 200 million years ago by the accretion of the southern continents, Australia, Antarctica, South America, India and Africa; began to rift apart 180 million years ago.

gossan   iron-rich weathered cap above a sulphide ore deposit, formed by leaching out of metals and leaving a rusty crust.

graben   rift valley formed by downfaulting of crustal block between two parallel normal faults (German: furrow, trough or trench).

grade of metamorphism   intensity of metamorphism, indicating general increase in pressure and temperature; e.g. shale to slate is described as low grade, while slate to schist or schist to gneiss is said to be high grade.

graded bedding   structure in sedimentary rocks (usually deposited in water), with coarser material at the base of a bed, gradually becoming finer towards the top.

Grampian Group   lowest group in Dalradian sequence, deposited as sands 9km thick, about 730 million years ago; beneath Appin Group; divided into Glenshirra, Corrieyarack and Glen Spean subgroups.

Grampian orogeny   deformation and metamorphism of Moine and Dalradian rocks in the Highlands, 480–465 million years ago; caused by collision of edge of Laurentia with an island arc; early part of the Caledonian orogeny (cf the later Scandian orogeny).

granite   coarse-grained acidic igneous rock, intruded as plutons into folded country rocks; contains quartz, plagioclase feldspar (sodium-rich) and orthoclase feldspar (potassium-rich), in nearly equal amounts, often also biotite mica; pink, white, grey or silver overall.

granular   rock texture in which mineral grains are of even size.

granulite   coarse-grained regional metamorphic rock formed under very high pressure and temperature conditions deep in the crust; granular texture, lacks banding or foliation; common in the Lewisian Complex, especially around Scourie and Gairloch; contains pyroxene, plagioclase feldspar, garnet, quartz.

greenstone   informal name for a metamorphosed basic igneous rock; green colour is due to alteration of hornblende etc. to chlorite, epidote and actinolite (an amphibole).

Grenville orogeny   a series of mountain-building events (folding, metamorphism, migmatite formation) in the Precambrian that led to the formation of the Rodinia supercontinent, comprising Laurentia, Baltica, Siberia and Gondwana, 1100–900 million years ago, and affecting the Moine rocks of the Northern Highlands;

eclogites formed in Glenelg.

greywacke medium-grained sedimentary rock, a type of sandstone, containing fragments of igneous and metamorphic rocks, angular quartz and clay minerals; formed by slumping of sediment on the continental shelf, then carried to the deep ocean by turbidity currents or mud slurries; grey, black, purple or dark green, tough and brittle, weakly metamorphosed.

grit coarse, angular sedimentary rock, usually referring to metamorphic quartzite in the Dalradian; preferred term is microconglomerate or metaconglomerate.

groundmass tiny fine-grained crystals in an igneous rock that also contains larger crystals (known as phenocrysts); also called matrix.

H

half graben an elongate sedimentary basin bounded by a normal fault on one side only.

hanging valley glacial landform feature caused by small tributary stream falling into a main valley that has been deepened by ice.

Hebridean Igneous Province Scottish part of the **North Atlantic Igneous Province**, active mainly 60–55 million years ago; previously known as the British Tertiary Igneous Province.

Highland Border Complex series of sedimentary and igneous rocks, including ocean-floor pillow lavas, of late Precambrian to Ordovician age (600–450 million years old) found patchily in a narrow belt along the Highland Boundary Fault.

Highland Border Downbend overfold of the Tay Nappe along the Highland Border, to form the Highland Border Steep Belt.

Highland Border Ophiolite part of the Highland Border Complex lying tectonically above the sedimentary Trossachs Group (top of the Dalradian); probably originated as slices of the mantle and ocean floor.

hinge part of a fold where two limbs meet and the fold curves over to form a line.

hornblende calcium–iron–magnesium–aluminium silicate, the commonest of the amphibole family; black to dark green elongate crystals, found in hornblende schist, amphibolite and banded gneiss.

hornfels a fine-grained crystalline rock found close to an igneous intrusion; results from baking of country rock; usually very hard and splintery.

I

Iapetus Ocean wide (up to 2000km) ocean between Laurentia (North America, Greenland and Scotland), Gondwana (Europe and the southern continents) and Baltica (Scandinavia and Russia) that existed from 580 to 420 million years ago; the line of collision between Scotland and England along the Solway Firth is known as the Iapetus Suture.

igneous rock rock formed by cooling and crystallization of molten magma in underground intrusions or on the surface as lavas.

ignimbrite igneous rock formed by ash and broken crystals in a violent volcanic eruption, so hot that clouds of material are welded together.

Glossary

imbricate overlapping packets of rock in a thrust belt, separated by small reverse faults.

inherited zircons zircon crystals found in metasedimentary rocks, but originally derived from an older source rock.

inlier older rock surrounded by younger.

inselberg landscape feature formed by steep, isolated mountain rising above low, flat terrain.

intrusion igneous body forced into country rocks at depth; minor intrusions are dykes (vertical), sills (parallel to beds, usually horizontal), cone sheets; major intrusions are granite and gabbro plutons.

intrusive igneous rocks rocks formed by physical intrusion or emplacement at depth.

Inverian folding, metamorphism and shearing events in the Lewisian Complex around Scourie and Lochinver, younger than the Badcallian (early Scourian) events.

island arc arc-shaped chain of volcanic islands formed when one oceanic plate is subducted beneath another at a **destructive plate boundary**; also known as a **volcanic arc**.

J

jasper bright red, hard form of silica (chert), rich in iron; found in Highland Border Complex.

joint natural crack or fracture in a rock, along which there has been no movement; cooling joints are common in igneous rocks.

jökulhlaup Icelandic name for catastrophic flood caused by sudden melting of ice.

Jurassic Period division of the geological column, 200–146 million years; above Triassic and below Cretaceous; named after Jura Mountains in France and Switzerland; North Sea oil deposits occur in Jurassic rocks; also found in Skye and Mull.

K

karst landform in limestone area; includes caves, dry river valleys, swallow holes, underground streams and limestone pavements; caused by carbonate being dissolved by rainwater.

knock and lochan topography formed in Lewisian Gneiss by glacial erosion of the landscape; low, bare, rounded rocky hills (cnoc in Gaelic) with small, shallow lochs and peat bogs between.

Knoydartian orogeny folding and metamorphism events in Moine rocks of the Northern Highlands 830–725 million years ago.

L

Laurentia Precambrian continent, consisting of North America, Greenland and Northwest Scotland; collided with Baltica and Avalonia to create the Caledonian mountain belt.

lava   stream of molten rock erupted onto the surface by a volcano (from a crater or a fissure), to form a lava flow.

Laxfordian   series of orogenic (mountain-building) events – folding, metamorphism, shearing and intrusion of granites and pegmatites – affecting Lewisian rocks 1900–1750 million years ago, to produce gneisses and migmatites; named after Loch Laxford in Sutherland.

layered intrusion   large basic to ultrabasic igneous intrusion in which crystals have settled into layers, under gravity, during cooling and crystallization; each layer usually has a distinctive composition, possibly including ore-mineral horizons.

Lewisian   Precambrian gneisses (3100–1750 million years old) forming the basement rocks of the Northwest Highlands and Outer Hebrides; also known as the Lewisian Gneiss Complex.

lineation   parallel alignment of minerals on cleavage or foliation surface of metamorphic rocks, often indicating direction of transport (stretching lineation).

Loch Eil Group   youngest of three divisions of the Moine in the Northern Highlands, mostly quartzites (called 'psammites'); 900–870 million years old; above Glenfinnan Group.

Loch Lomond Readvance (Loch Lomond Stadial)   cold period at the end of the most recent ice age, 11,000–10,000 years ago; re-establishment of a mountain icecap from Torridon to Loch Lomond, after the melting of the main ice sheet.

Loch Maree Group   sequence of metamorphosed igneous and sedimentary rocks in the Lewisian Complex, 2000 million years old.

M

machair   coastal landform in western Scotland, consisting of shell sand covered by rich, fertile soil; sand dunes and white sandy beaches at sea level.

mafic   igneous rock containing dark-coloured silicates (olivine, pyroxene, amphibole) and feldspar; contrasts with felsic.

magma   molten rock, usually containing dissolved gases; cools to form crystalline igneous rocks.

magma chamber   body of magma in the crust which feeds lava to surface volcanoes; crystals forming in the chamber may become stratified to form a layered intrusion.

magnetite   magnetic iron oxide; cubic mineral and important iron ore; very common as a minor mineral in basic igneous rocks.

mantle   layer of the Earth below the crust and above the core; made predominantly of peridotite; upper mantle contains the weak, partially molten asthenosphere; boundary between crust and mantle is marked by the Moho.

marble   metamorphosed limestone (meta-limestone) containing calcite (calcium carbonate), dolomite (calcium–magnesium carbonate) and sometimes other minerals; has sugary or granular texture.

marker horizon   a layer of rock that has certain distinctive features (e.g. composition, fossils) that allow it to be used over a wide area as a point of reference; particularly useful in mapping.

Mesozoic Era   division of the geological column, from 252 to 66 million years ago; includes Triassic, Jurassic and Cretaceous periods; follows Palaeozoic and precedes Cenozoic eras; from Greek for 'middle life'.

metamorphic grade   relative intensity of pressure and temperature conditions during metamorphism; slate is a low-grade rock, schist is medium-grade, gneiss is high-grade.

metasediments   informal term for metamorphosed sedimentary rocks.

mica   sheet-like complex potassium–aluminium silicate mineral with good cleavage; muscovite is colourless, biotite is black or dark brown (because of iron content); muscovite is common in schist, biotite is common in granite, schist, gneiss and amphibolite.

migmatite   high-grade metamorphic rock of mixed composition (usually granite and amphibolite) and patchy appearance, resulting from partial melting and injection of granitic veins into more basic rock; very common in the Lewisian Gneiss Complex. (Greek: mixed rock).

Moine Thrust zone   narrow belt of faulted rocks in Northwest Scotland marking the edge of the Caledonian fold belt, where Moine rocks were transported 430 million years ago at least 100km to the northwest across the foreland of Lewisian, Torridonian and Cambrian–Ordovician rocks; characterized by overlapping slices of rock, faulted and piled one on top of the other; thick mylonite zones are frequent.

Moine   Precambrian metamorphosed sediments (schist and quartzite), 1000–870 million years old, north of the Great Glen Fault and east of the Moine Thrust zone in the Northern Highlands; also called Moine Supergroup.

moraine   mound of unstratified and unsorted glacial **drift** deposited by a glacier; terminal moraine forms at the front or snout of a glacier, lateral moraine along valley walls, and medial moraine where two glacier flows meet and their lateral moraines merge.

Morar Group   oldest unit of Moine rocks, deposited as sediments about 980 million years ago; lies above the Moine Thrust and separated from the Glenfinnan Group by the Sgùrr Beag Thrust. May correlate with Torridon Group.

Morarian   folding and metamorphism of Moine rocks 750–730 million years ago.

mudcracks   structure formed in fine sediment when ponds or river beds dry up and mud or clay dries and shrinks; useful way-up indicator.

mullion   structure in folded quartzites resembling parallel cylinders or large rods.

muscovite   white mica, a sheet silicate; splits easily into very thin flakes; common in schist.

mylonite   fine-grained metamorphic rock with banded structure, from extreme crushing and recrystallization of pre-existing rocks in narrow shear zones and thrusts; from Greek word for 'mill'.

N

nappe   large sheet of rock transported above a thrust fault; from French for 'tablecloth'; nappe structures are common in the Northwest Highlands, in the Moine

Thrust Zone.

New Red Sandstone Permian and Triassic rocks, deposited in desert conditions.

normal fault fault with downward movement of one side relative to the other.

North Atlantic Igneous Province broad area of igneous activity during the Palaeogene, 60–55 million years ago, embracing East Greenland and Northwest Scotland that produced large volumes of basalt lava from central complexes, e.g. Mull, Skye, Rum, Ardnamurchan; may have resulted from a mantle plume intruding the base of the crust.

O

Old Red Sandstone continental sedimentary rocks formed in valleys and floodplains during mainly the Devonian Period by rivers flowing off the Caledonian Mountains; the red colour is due to iron staining in the cement.

olivine iron–magnesium silicate mineral found in basic and ultrabasic igneous rocks; dark green or black and very dense, weathers easily at the surface; alters to serpentine during metamorphism.

ophiolite slice of old oceanic crust (basalt) thrust into folded continental rocks during mountain building; often metamorphosed to serpentinite.

Ordovician Period division of the geological column, 488–444 million years ago, above Cambrian and below Silurian; named after the ancient Welsh tribe, the Ordovices.

orogeny mountain-building episode, a sequence of events including folding, metamorphism, thrusting, faulting, granite intrusion and crustal thickening, occupying several tens of millions of years.

outlier isolated body of younger rock surrounded by older rocks and situated at a distance from the main outcrop.

P

Palaeozoic Era division of the geological column, from 542 to 252 million years ago; includes Cambrian, Ordovician, Silurian, Devonian, Carboniferous and Permian periods; follows Precambrian, and precedes Mesozoic era; from Greek for 'old life'.

partial melting process in the upper mantle and lower crust that generates magma; pressure at depth is too great to allow complete melting, but if pressure is slightly released, and small amounts of water are present (1–5%), then peridotite can melt and produce basalt lava; in the case of thick sediments in a mountain chain, high temperatures allow granite to form by partial melting of the lower crust; melts are lighter than the surrounding rocks and so rise towards the surface; partial melting is a key stage in the rock cycle.

pebbles rounded rock fragments, 4–64mm in diameter; sedimentary rock containing pebbles is known as conglomerate.

pegmatite extremely coarse-grained igneous rock, usually granite in composition; contains large, well-formed crystals of quartz, pink feldspar and mica over 25mm in size; formed in the final stages of cooling and crystallization of magma containing water-rich fluids.

pelite   term used to describe slates and schists in Moine and Dalradian metasedimentary units; originally they would have been shales, siltstones and mudstones; contrast with psammite.

peridotite   very coarse-grained dense, black ultrabasic igneous rock containing olivine and pyroxene (iron–magnesium silicates); the material that makes up the mantle; also found in large layered igneous intrusions; crystals may have settled out of the magma to form separate layers; may be altered to serpentinite during metamorphism in the presence of water-rich fluids.

Permian Period   division of geological timescale, 299–251 million years; above Carboniferous and below Triassic; named after Perm in eastern Russia; Permian and Triassic desert sandstones are named the New Red Sandstone in Britain.

phenocryst   large crystal in an igneous rock, surrounded by finer groundmass.

phyllite   green-coloured low-grade metamorphic rock (between slate and schist in grade), typically with closely spaced tiny folds, crinkles and corrugations (crenulations), giving a wavy appearance; phyllite splits easily but irregularly; green colour derives from the presence of chlorite, a silicate with a sheet structure resembling mica.

pillow lava   form of basalt erupted from submarine volcanic vents and chilled to form an outer fine-grained skin.

Pipe Rock   distinctive pinkish Cambrian quartzite with vertical worm tubes cutting across bedding planes; an important marker horizon for stratigraphy and structural geology in the Northwest Highlands.

plagioclase   group of the feldspar silicate minerals, containing calcium and sodium; the commonest minerals in igneous and metamorphic rocks.

plate tectonics   theory of global tectonics that states that the lithosphere is divided into a number of large rigid plates, which are capable of moving across the surface of the globe and which interact at their boundaries.

plateau lavas   thick series of lava flows, usually basalts (and known as **flood basalts**), forming an extensive flat-topped plateau across thousands of square kilometres.

porphyritic   said of igneous rocks with large crystals set in a fine groundmass (or matrix).

post-orogenic   igneous rocks formed shortly after the end of a mountain-building episode; they are therefore undeformed.

Precambrian   division of geological time, older than Cambrian, 542 million years; generally devoid of fossils with hard parts; usually subdivided into Hadean (4560 to 4000 million years), Archaean (4000–2500 million years) and Proterozoic (2500–542 million years); followed by the Palaeozoic Era.

pressure solution   process in rocks undergoing deformation by which edges of mineral grains dissolve under high stress, along contacts with neighbouring grains, in the presence of water-rich fluid; the dissolved material is removed and deposited elsewhere within the rock.

protolith   original rock from which a metamorphic rock is derived.

psammite   metamorphosed sandstone (quartzite); used mainly in older descriptions of Moine and Dalradian rocks; contrasts with pelite.

pyrite iron sulphide, FeS_2, a brassy yellow iron ore that forms cubes; also known as fool's gold; common in slate; may form nodules in sedimentary rocks; also called iron pyrites.

pyroclastic volcanic rock formed of broken crystals, rock and ash during explosive eruptions.

pyroxene family of calcium–iron–magnesium silicates; dense, dark in colour; very common in basic and ultrabasic igneous rocks; augite is one of the commonest pyroxenes.

Q

quartz silicon dioxide, SiO_2, one of the commonest rock-forming minerals, especially in granite; sandstone consists mostly of quartz grains; harder than glass; colourless or white; perfect hexagonal crystals are known as rock crystal.

quartzite metamorphosed sandstone in which quartz grains and silica cement have recrystallized to form a rock with a granular texture in which the grains are evenly shaped; very hard, white and tough rock; the Cambrian quartzite of Northwest Scotland is a form of sedimentary quartzite with a recrystallized silica cement.

quartzo-feldspathic used of metamorphic rocks made of quartz and feldspar, e.g. granite gneiss.

Quaternary informal division of the Cenozoic, from 1.8 million years ago until the present. Includes the various glacial and interglacial phases of the climate, when considerable erosion of the surface took place, resulting largely in the present land surface of the Highlands. Deposits include moraines, till ('boulder clay') and drift.

R

recumbent fold a fold that has been overturned so that its axial plane is horizontal, i.e. a flat-lying fold as opposed to an upright fold.

regional metamorphism wholesale alteration of rocks over a wide area, involving increases in pressure, temperature and fluid activity and resulting in recrystallization during deformation within a continental mountain chain.

retrograde type of metamorphism that results from reheating and water-rich fluid effects in rocks previously metamorphosed at a higher grade; opposite of prograde.

reverse fault steep fault in which one block of rock is pushed up over another.

rhyolite a fine-grained acid lava, usually grey or white in colour and rich in quartz and feldspar; the surface equivalent of granite.

rift structure in the crust resulting from stretching and downfaulting along a pair of parallel faults; also known as a graben.

ring dyke minor igneous intrusion in the form of a vertical cylindrical wall.

ripple marks marks left on the surface of sediment by currents washing back and forth; a useful way-up indicator.

roche moutonnée a landform created by glacial erosion when bedrock outcrops are streamlined by moving ice that climbs up a smooth whaleback and plucks material away from the sheltered steep side (French: refers to a wig held on the head by pieces of sheep's fat).

rodding   type of lineation in metamorphic rocks where components have been stretched into parallel rods; cf mullion structure.

Rodinia   supercontinent that formed 1100 million years ago by the collision of Laurentia, Baltica, Siberia and Gondwana (i.e. nearly all the continents) in the Grenville orogeny, when parts of the Northern Highlands formed; Rodinia began to rift apart 750–600 million years ago, when the Iapetus Ocean formed as a result; Rodinia is derived from the Russian for 'motherland'.

S

Scandian orogeny   folding, metamorphism and thrusting events in Northern Highlands 435–425 million years ago; late stage of Caledonian orogeny; cf earlier **Grampian orogeny**.

schist   medium- to coarse-grained metamorphic rock rich in mica, foliated and roughly banded, result of high-grade **regional metamorphism**; the main or significant mineral is often attached, e.g. mica schist, garnet schist, hornblende schist.

schistosity   foliation structure in schist due to parallel alignment of platy mica.

Scourian   folding and metamorphism of the early crust in the Northwest Highlands, to produce Lewisian gneiss and granulite, 2900 million years ago; now divided into earlier Badcallian and later Inverian events; named after Scourie in Sutherland.

Scourie Dykes   swarms of northwest–southeast dolerite dykes intruded into Lewisian gneiss 2400 million years ago.

scree   loose angular material lying at a steep angle on a hill slope, caused by weathering and collapsing of bedrock; accepted term is now talus.

sedimentary basin   large flat depression in the continental crust into which rivers flow and deposit their sediment load; usually long-lived and with thick sequences of sedimentary rocks resulting from continuous subsidence of the basin floor; may be bounded by faults that were active during sedimentation.

sedimentary breccia   coarse angular sedimentary rock formed usually by the rapid removal and deposition of rock fragments close to the source area.

sedimentary rock   rock formed on the Earth's surface by processes of weathering, erosion, transportation, deposition, burial, compaction and cementation (clastic or fragmental rocks); or by biological activity such as reef-building animals to create limestones; or as chemical precipitates, such as rock salt (halite, by evaporation of sea water).

serpentine   group of hydrated iron–magnesium silicate minerals with a sheet structure; usually green and may be fibrous; result from metamorphism or hydrothermal alteration of olivine and pyroxene, sometimes also amphibole.

serpentinite   metamorphic rock made mainly of serpentine group minerals, often also with chlorite, carbonate minerals, talc and magnetite; formed from olivine and pyroxene in basic and ultrabasic igneous rocks; dark green, streaked with black and red, greasy feel.

Sgùrr Beag Thrust a major shear zone within the Moine nappe in the West Highlands, separating Glenfinnan and Morar groups of the Moine rocks.

shale thinly bedded fine-grained sedimentary rock rich in clay minerals and fine sand.

shear deformation caused by one part of a body of rock sliding sideways against another.

shear zone narrow zone of highly deformed rock in which recrystallized minerals are often strongly aligned; important for the movement of fluids in the crust and may be the sites of ore deposits.

shelf sea shallow sea above the continental shelf, usually less than 250m deep; the North Sea is a shelf sea.

shield large stable ancient area of Precambrian rocks at the centres of the major continents; lacking in earthquake and volcanic activity and usually surrounded by linear mountain chains.

sill sheet of magma intruded into sedimentary rocks parallel to bedding planes.

Silurian Period division of geological timescale, 444–416 million years, above Ordovician and below Devonian; named after the ancient Welsh tribe, the Silures.

sinistral left-directed movement of a fault or shear.

sink hole feature in limestone areas where a stream disappears underground, due to solution of the bedrock by rainwater; also known as swallow hole.

slate fine-grained low-grade metamorphic rock rich in platy minerals and with a pronounced cleavage that allows it to be split into thin slabs.

slaty cleavage structure in slate caused by parallel alignment of platy minerals such as mica and chlorite.

Sleat Group lowest part of the Torridonian, found in southeast Skye and Kyle of Lochalsh; sandstones and shales laid down by rivers and lakes; very weakly metamorphosed.

slickensides parallel grooves and linear scratch marks on a fault plane caused by slippage and grinding of one block against another.

Southern Highland Group sequence of 5km of metasedimentary and meta-igneous rocks towards the top of the Dalradian; above the Appin and below the Trossachs groups.

Stac Fada Member distinctive marker bed in the Stoer Group (Torridonian), with ash pellets and broken crystals, possibly resulting from a meteorite impact.

Steep Belt vertical or overturned Dalradian rocks along the Highland Border, within the Highland Border Downbend.

Stoer Group early part of the Torridonian, 1200 million years old, below the main Torridon Group, lying directly on Lewisian basement and formed of coarse conglomerate and sandstone; divided into Clachtoll, Bay of Stoer and Meall Dearg formations; named after the Stoer Peninsula in northwest Sutherland.

striations (striae) scratches or parallel marks caused by glaciers (glacial striae) or during faulting (slickensides).

strike direction or trend of a structure, at right angles to the dip direction of a bed.

strike-slip fault fault in which the movement is sideways, parallel to the strike of the fault; e.g. the Great Glen Fault; sometimes called **transcurrent fault**, tear fault or wrench fault; blocks move relatively to the right (dextral fault) or left (sinistral fault).

stromatolite   pillow-shaped or mushroom-shaped mounds of algal growth forming large carbonate reef-like structures in shallow water; the oldest known fossils belong to this group.

subduction   tectonic process whereby an oceanic lithospheric plate descends below another plate into the upper mantle.

subduction zone   long, narrow belt at a destructive plate margin where subduction takes place; marked by an ocean trench; accompanied by shallow and deep earthquakes.

subsidence   sinking or settling of the Earth's surface; may be caused by cooling and relaxation of crust after volcanic activity (thermal subsidence); or collapse of a large volcanic structure around ring faults (cauldron subsidence); or collapse due to subsurface workings (mining subsidence).

supercontinent   amalgamation of a large group of continents into one unit, e.g. Pangaea.

suture   a line marking the join of two separate plates at a continent–continent collision zone.

swallow hole (sink hole)   landform feature in limestone country, where a surface stream disappears underground due to dissolution of limestone by rainwater.

syenite   coarse-grained intrusive igneous rock, similar to granite, but very low in quartz; rich in feldspar, and usually deep red in colour; forms in continental crust by small amount of partial melting of an igneous parent rock.

syncline   a V-shaped fold in which the two limbs slope in towards each other, so that younger rocks are in the centre or core of the fold; opposite of an anticline.

T

talus   loose angular material forming a sheet lying at a steep angle at the foot of a hill or cliff, caused by weathering and collapsing of bedrock; cf scree.

Tay Nappe   large overfold in the Dalradian, running parallel to and along the entire length of the Highland Border, consisting of a Flat Belt to the north and a Steep Belt adjacent to the Highland Boundary Fault.

terrane   large segment of the crust, bounded by major strike-slip faults, in which the age, composition and geological evolution of the rocks are distinct from adjacent terranes.

Tertiary   informal division of geological time, 66–2.6 million years; above Cretaceous and below Quaternary; now replaced by Palaeogene and Neogene.

texture   the way in which the grains, particles or crystals in a rock are held together, and relate to each other; e.g. crystalline texture, fragmental texture.

thermal metamorphism   process of recrystallization of country rocks around a large igneous intrusion in a thermal aureole (metamorphic aureole), due to heat exchange.

throw   the amount of movement or displacement along a fault.

thrust fault   reverse fault with movement plane shallower than 45°; horizontal compression forces one block across another.

till   loose glacial drift, unsorted, unstratified, with mixed sizes and compositions of the material; formed by deposition under an ice sheet; often referred to as 'boulder clay'.

tillite   sedimentary rock made of fossilized glacial till.

tombolo   narrow spit of land, usually sand and gravel ridge, joining a small island to the mainland.

Torridon Group   red sandstone, arkose and conglomerate, 1000–800 million years old, forming imposing mountains on the northwest seaboard; lies above Stoer Group; divided into Diabaig, Applecross and Aultbea formations.

Torridonian   informal name for the Torridonian Supergroup, an important sedimentary formation lying above the Lewisian gneiss basement and below Cambrian–Ordovician sediments of the Northwest Highlands, west of the Moine Thrust Zone.

trace fossil   structure in a sedimentary rock created by an animal whose remains are not preserved, e.g. worm tubes, burrows, feeding trails, footprints.

transcurrent fault   a strike-slip fault (wrench fault) in which blocks on opposite sides of a vertical fault plane move sideways past one another.

trap topography   a step-like landscape feature formed by horizontal basalt lavas, usually with a fossil soil horizon (bole) between each flow, which is preferentially weathered out; typically seen in the Inner Hebrides and on Ardnamurchan.

Triassic Period   division of geological time from 251 to 200 million years ago, above Permian and below Jurassic. Permian and Triassic together are informally called the New Red Sandstone. Rocks of this age form patchy outcrops of river-deposited sandstones and conglomerates around the north and west coasts.

trilobite   extinct fossil arthropod, with head, segmented body and tail; lived from Cambrian to Permian.

Trossachs Group   the youngest part of the Dalradian, above the Southern Highlands Group; sedimentary part of the Highland Border Complex, beneath the Highland Border Ophiolite; age is early Cambrian to early Ordovician.

TTG gneiss   tonalite–trondhjemite–granodiorite igneous composition (roughly granitic) of the early Lewisian crust; often pale grey to creamy, without pronounced banding; also referred to as 'grey gneiss'.

tuff   consolidated volcanic ash, often bedded.

U

ultrabasic igneous rock   igneous rocks with less than 44% silica; made almost entirely of olivine and pyroxene, plus some iron ore, i.e. the ferromagnesian silicates, with no feldspars and no quartz; dense, black rocks, usually very coarse grained, found in large intrusions, sometimes with crystals in layers (i.e. layered intrusions), e.g. peridotite; also known as ultramafic rocks.

unconformity   a break in the geological record, sometimes expressed as younger rocks lying on top of older rocks that have been uplifted and folded; represents a time gap.

upthrow   side of a fault that has moved upwards relative to the opposite side.

V

volcanic arc   arc-shaped chain of volcanic islands formed when one oceanic plate is subducted beneath another at a destructive plate boundary; also known as an island arc.

W, X

way-up structure   feature in a sedimentary rock or metasediment that indicates the top and bottom of the beds, e.g. graded bedding, cross bedding, ripple marks, mud cracks.

welded tuff   a fine-grained volcanic rock resulting from the compaction of ash and crystal shards due to the intense heat in a burning, explosive cloud.

xenolith   block of country rock enclosed (caught up in) an igneous rock; Greek for 'foreign rock'.

Z

zinc blende (sphalerite)   zinc sulphide, ZnS, a brown cubic mineral, an ore of zinc, often found with galena (lead ore), e.g. at the Strontian and Tyndrum mines in Argyllshire.

zircon   common mineral in granite and gneiss; colourless silicate of zirconium, forming tiny elongate crystals, often inside biotite mica; has very high melting temperature and forms early in the crystallization of igneous rocks; stable and resistant mineral; often contains small amounts of radioactive uranium and hence is useful for age-dating purposes (see also inherited zircons and detrital zircons).

Gaelic terms

abhainn   river
ach, achadh, auch   field
allt   stream
àrd   high, or headland
avon   anglicized form of abhainn

bad   cluster of houses; or a hollow
bàgh   bay
baile   township, scattered settlement, town
bàn   white
beag, bheag   small
bealach   mountain pass (pronounced 'byalach')
beinne   mountain
ben   anglicized form of beinne
binnein   pinnacle, hill summit
bò   cow (plural **bà** cattle)
bog   soft
buidhe   yellow

cairn   anglicized form of càrn
càm   bent, crooked
camas   bay, inlet, beach
caol   narrow (usually inlet of the sea)
càrn   heap of stones
ceann   head (pronounced 'kyawn')
ceum   step
cille   chapel, monk's cell
cìr   comb, crest of a hill
clach   stone
cnoc   rounded hill (pronounced 'krok')
cnocan   small hillock
còinich   mossy
coire   corrie (glacial landform; originally means pot or cauldron)
crag   anglicized form of creag

Gaelic terms

creag   stone, rock, cliff
cruach, cruachan   pile of stones
cùl   round hill (shaped like a person's back)

dearg   red (pronounced 'jerrak')
drum, druim   ridge
dubh   black
dùn   prominent hill, fort

eas   waterfall
eilean   island

fada   long
fiacal   tooth

garbh   rough (pronounced 'garav')
geall   white
glas   greyish green
gleann   narrow valley (pronounced 'glyawn')
glen   anglicized form of gleann
gorm   green, bluish green

inbhir   mouth of a river
inch, insh   anglicized form of innis
innis   pasture beside a river, or island in a river
inver   anglicized form of inbhir

kil   anglicized form of cille
kin   anglicized form of ceann
knock   anglicized form of cnoc
kyle   anglicized form of caol

lairig   high mountain pass
liath   grey
loch   lake
lochan   small lake

machair   flat field beside the sea
maol   bare, round hill
meadhon   middle
meall   shapeless hill
mòine   peat, peat bog
monadh   mountain (pronounced 'mona')

mòr, mhòr   large
mullach   headland

nis   promontory (from Norse ness)

òb   bay
òrd   high place or round hill

poll   pool
port   harbour, bay

rannoch   bracken
rinn   point of land, promontory, headland
ros, ross   promontory or round hill
ruadh   red
rubha   promontory, peninsula

sàil   heel, end of a hill ridge
sgùrr, sgòrr   peak, rocky summit
sithean   little round hill, fairy knoll (pronounced 'she-an')
spidean   sharp peak, pinnacle
srath, strath   wide river valley
sròn   nose (pronounced 'stron')
sruth   stream (pronounced 'stroo')
stac   steep conical hill
stòr, stòrr   steep high cliff, pinnacle
stron   anglicized form of sròn

taigh   house
tìr   land (pronounced 'cheer')
toll   hollow (pronounced 'towl')
tom   small hill
tòrr   round hill
tràigh   shore, beach

uamh   cave (pronounced 'ooav')
uig   bay (from Norse vík)
uisge   water (pronounced 'ooshki')

Index

Page numbers in *italic* denote figures. Page numbers in **bold** denote tables.

Index